PRAISE FOR *EXPAND*

"As global challenges are accelerating, public servants and policy makers need more than new tools; they need a new way of thinking. *Expand* offers just that: a fresh, ambitious perspective on how to innovate in the service of society and the planet. The authors provide a compelling account of how we must build on the foundations of design thinking to embrace a new class of problem-solving—to the benefit of public institutions and the cities, nations, and people they serve."
—Lisa Witter, Cofounder and CEO, Apolitical

"The last century saw the role of design in society grow from enabler of industrialization and mass production to drivers of digital services and systems design. In this book, Jens Martin Skibsted and Christian Bason contend that now is the time to widen the design discourse to take on the grand, complex challenges of tomorrow. With a uniquely Danish perspective, they examine the universal power of design for shaping a society and call on designers, businesses, and leaders alike to broaden their ideas about 'design thinking' and adopt new ways to use design to create wider value for generations to come."
—Mariana Mazzucato, Professor, Economics of Innovation
and Public Value, University College London

"*Expand* lays out an ambitious path to unleashing the full power of design thinking as a key approach to innovation and sustainable growth. If we are to effectively address the urgent challenge of climate change, we must design more circular value chains and business models. This book provides the vision, the thinking, and the cases we need to get started. A must-read for anyone seeking to positively impact our common future."
—Jim Hagemann Snabe, Chairman, Siemens and AP Moeller-Maersk, and former Co-CEO, SAP

"All too few methods, theories, insights, and approaches make themselves truly available to collective problem-solving at scale. All too often that kind of inspiration stays evasive. Here's a bold attempt to fully engage with the biggest challenges we face. It's highly ambitious, clearly argued—and profoundly thought-provoking".
—Bjarne Corydon, CEO and Editor-in-Chief, *Borsen Daily*,
and former Minister of Finance of Denmark

"*Expand* is an important eye-opener that stretches our thinking about innovation. This timely book shows us how we meet global challenges in an increasingly complex world. It's a must-read for entrepreneurs, business managers, and public servants—leaders at all levels—who are working to create a better world."
—Alan Webber, Mayor of Santa Fe, New Mexico, and founder of *Fast Company*

"This book is a major contribution to the field of innovation in the 21st century. It cements that technology is not an exogenous force over which we have no control. Technological change should lead us to reflect about who we are and how we see the world. The insight that new solutions can emerge from anywhere—across geographies, technologies, and cultures—challenges long-held assumptions that innovation stems from a specific linear and incremental engineering tradition, often too absorbed by immediate concerns. *Expand* is a must-read for leaders and decision-makers wishing to embrace the promise of new technology and improve the state of the world."
—Klaus Schwab, Founder and Executive Chairman of the
World Economic Forum and the Davos Agenda

"Jens Martin Skibsted and Christian Bason in their new book speak to a land that I have been in search for my entire design life. A land that was first discovered by Lou Kahn, the great architect. A land called Zero. A land of beginnings, of a parallel universe in which everything is designed. The design of your life, the design of a chair, the design of humor, the design of music. It's not the business of design but the design of business. They strike out a clear path that, in its simplicity of expanding our mindset about design, is remarkable."
—Richard Saul Wurman, Cofounder of the TED conference

"In *Expand*, you'll learn a revolutionary approach to design that works for any situation. By offering a new way to think about innovation and design, it gives all of us—not just designers—the blueprints to solve problems and take advantage of opportunities."
—Mark Frauenfelder, Research Director, Institute for the Future,
and former Editor-in-Chief of *Make* magazine

EXPAND

**ALSO BY CHRISTIAN BASON &
JENS MARTIN SKIBSTED**

By Christian Bason

Leading Public Sector Innovation: Co-creating for a Better Society
(2nd ed.). Policy Press (2018)
Leading Public Design: Discovering Human-Centred Governance.
Policy Press (2017)
Design for Policy (ed.) Routledge (2014)

By Jens Martin Skibsted

Tilbage til virkeligheden: Sådan skabes værdi i en verden, hvor alle ved alt. With Rasmus Bech Hansen. Gyldendal (2013)
Instant icon: Om produkter der skaber exceptionel værdi og hvordan de bliver til. Gyldendal (2008)

EXPAND

STRETCHING THE FUTURE BY DESIGN

JENS MARTIN SKIBSTED & CHRISTIAN BASON

Matt Holt Books
An Imprint of BenBella Books, Inc.
Dallas, TX

Expand copyright © 2022 by Jens Martin Skibsted and Christian Bason

All rights reserved. No part of this book may be used or reproduced in any manner whatsoever without written permission of the publisher, except in the case of brief quotations embodied in critical articles or reviews.

Matt Holt is an imprint of BenBella Books, Inc.
10440 N. Central Expressway
Suite 800
Dallas, TX 75231
benbellabooks.com
Send feedback to feedback@benbellabooks.com.

BenBella and *Matt Holt* are federally registered trademarks.

Printed in the United States of America
10 9 8 7 6 5 4 3 2 1

Library of Congress Control Number: 2021057359
ISBN 9781637740736 (trade cloth)
ISBN 9781637740743 (ebook)

Editing by Katie Dickman
Copyediting by Ginny Glass
Proofreading by Marissa Wold Uhrina and Ashley Casteel
Indexing by WordCo Indexing Services, Inc.
Text design and composition by Aaron Edmiston
Cover design by Brigid Pearson
Cover image © Shutterstock / Stilesta
Printed by Lake Book Manufacturing

Special discounts for bulk sales are available.
Please contact bulkorders@benbellabooks.com.

For our next generation, to whom the future belongs:
Olga, Mads Ludvig, Kai & Ella
Lillian, Julia & Christopher

CONTENTS

	Introduction	1
1	Design Past and Present	25
2	Expansion 1—Time	33
3	Expansion 2—Proximity	57
4	Expansion 3—Life	87
5	Expansion 4—Value	107
6	Expansion 5—Dimensions	129
7	Expansion 6—Sectors	151
8	Applying the Expansions	175
	Appendix	193
	Acknowledgments	201
	References	203
	Index	225

INTRODUCTION

> The ultimate, hidden truth of the world is that it is something we make, and could just as easily make differently.
> David Graeber, *The Utopia of Rules*

> A dream will always triumph over reality, once it is given the chance.
> Stanislaw Lem, *The Futurological Congress*

We've all been there. Stuck on hold for what feels like forever while the customer service department looks into your case for the umpteenth time. And the bigger the beast, the more byzantine the bureaucracy. Which is what makes recent events at the US Department of Veterans Affairs so interesting.

Few government departments generate as much goodwill as the VA. The largest civilian agency in the United States, it serves the needs of millions of military veterans and their families by providing lifelong health care, disability compensation, and other benefits and services. Its motto comes from President Lincoln's second inaugural address, written in the final days of the Civil

War: "To care for him who shall have borne the battle and for his widow, and his orphan."

So when news broke in 2014 that at least forty veterans had died waiting for appointments at a VA hospital in Phoenix, it threw the department into a tailspin. Just as galling, many of those veterans had been placed on a secret waiting list by VA managers trying to conceal the fact that hundreds of sick veterans had to wait months to see a doctor. As one VA official put it, the department experienced "an extreme pain point and crisis."

Other damaging headlines followed. In 2015, a Department of Veterans Affairs inspector general report revealed that three hundred thousand military veterans likely died while waiting for health care. And seeing a doctor wasn't the only thing America's soldiers had to wait for. The process for appealing benefits-claims decisions turned out to be similarly dysfunctional. Between 2012 and 2015, the number of pending appeals soared by 35 percent to over 450,000.

Inevitably, political pressure grew. In 2017, Congress passed the Veterans Appeals Improvement and Modernization Act, which updates the VA's benefits-claims appeals process and shortens the average wait time for an appeal from five years to 125 days. "This streamlined process would provide veterans with timely, accurate answers on their appeals so they can access the benefits they need and deserve," said Democratic Senator Richard Blumenthal when the bipartisan bill was introduced in May 2017.

But the interesting thing is what happened at the VA itself. You see, its tailspin came when a group of designers and technologists were exploring ways to improve American citizens' ability to access government services such as food stamps and health care. They included Sarah Brooks, a Presidential Innovation

INTRODUCTION

Fellow in 2014–15 and a designer who became the VA's first-ever design officer and led a team that launched the Veterans Experience Office. Its purpose? To revamp the way veterans interacted with the VA.

The department's digital services team began by examining the VA's assets—online forms, software, and so on—and talked to veterans to determine their pain points when interacting with the VA. Those included fundamental problems such as being unable to apply for benefits unless one had the correct version of Internet Explorer or Adobe Reader. "Accomplishing just about anything with the VA—scheduling a medical appointment or getting hired or filing a claim—whether online or in person was challenging, to put it mildly," wrote design writer Allison Arieff in the *New York Times*.

The design team also represented the VA visually, illustrating the scope of its bureaucracy in a rendering that looked more like a map of the London Underground than a functioning organization. It was almost certainly the first time that anyone at the VA had captured an overview of its size and sprawl. Visualizing the organization and its interactions with its users and external stakeholders in this manner was crucial—because if the VA couldn't see what was wrong, how could it hope to fix it?

More extraordinarily still, the team discovered that veterans had to navigate more than 450 websites to connect to VA services. In time, it consolidated these sites into an easy-to-navigate, service-delivery platform called Vets.gov—drawing inspiration from the one-stop-shop redesign of the British government's citizen-facing website, Gov.uk. "The notion of citizen as customer, rather than cog, drives these projects," explained Arieff in the *New York Times*. "A visitor to a governmental website should be treated like a customer to Amazon.com—the experience should be quick, convenient and intuitive, one that he or she would engage with again."

What happened at the VA shows us what happens when we take the principles of one field—design—and apply them to a seemingly unrelated one, in this case, the way that people access government services online. The VA's turnaround also demonstrates how far the principles of design have come. The purpose of this book is to demonstrate how much further they can go.

BRINGING DESIGN UP TO SPEED

The American industrial designer Charles Eames was once asked to define the boundaries of design. "What are the boundaries of problems?" he replied. That was more than fifty years ago. The impact of design has only grown since then. As the boundaries of problems have expanded, so has design. We believe it must expand even more in the years to come.

The problems facing business and indeed humanity are becoming even more intractable, intertwined, and complex. The global population is exploding to near ten billion by midcentury. Cities are mushrooming by some 2.5 billion people by 2050, making them more crowded and clogged. We need to build a city the size of New York every other month for the next thirty-five years in order to house the future global urban population. Meanwhile, the global demand for everyday goods, housing, and transportation is rising—requiring material resources that we do not have and squeezing our planet's ecosystems. Mass migrations, booming income inequality, raging pandemics, and the impact of AI and automation add dramatically to the challenges we face.

INTRODUCTION

The world is screaming out for new solutions to tackle the myriad problems and mitigate their consequences. This calls for innovation—bold, new approaches that can be adopted by people, businesses, public organizations, NGOs, and even international bodies.

This book makes the case for what we call *expansive thinking*. Its central premise is that we should expand the scale of design thinking and the scope of human-centered design. Make no mistake: if we want to create a better, more prosperous, and more sustainable world, we still need to design new products and services. But if we're serious about tackling the world's thorniest problems—from climate change and resource scarcity to income inequality and the impact of AI—we won't just need good design. Rather, we need to expand the very way we think as we seek to create solutions to our biggest challenges. In so many ways, however, the way we innovate today no longer reflects reality. We believe we can use a more ambitious set of design principles to better respond to the transformations that are occurring, and we hope to show you how.

Now, while many readers may recognize what happened at the VA as an example of human-centered design or the broader, management-oriented field of "design thinking"—which is to say, the application of design principles such as prototyping and empathizing with the people being designed for—rest assured, this isn't another book about those fields. While human-centered design (and its close cousin "user-centered design") has its roots in the Scandinavian tradition of cooperative design—making it ripe for discussion by two Danish writers—it is well-trodden ground. We want to broaden our horizons, go beyond the edge of the map, and explore terra incognita. If the overhaul of the VA shows us where design has taken us, this book seeks to blaze the trail ahead of us.

EXPAND

Straddling policy and technology, systems and users, design has today become a universal medium for expressing ideas, raising fundamental questions, and addressing societal challenges. After all, anyone can have a good idea—but it's worth very little until it's realized as, say, a website, a physical product, an app, or an organization.

So, design is everywhere now. Or as the American futurist R. Buckminster Fuller famously claimed, "Design is everything." The days when designers were primarily concerned with physical objects and graphical communication—or "toasters and posters"—are long gone. Design now embraces much wider areas of creative activity.

Design's ubiquity in our lives is one reason why we think it's the key to creating a better world. Another is that designers can think creatively across materials, methods, sectors, and professions. They don't just create attractive products or services; by devising new forms of organization and approaches to value creation, they can enable impactful change on a global scale. Design's applications are therefore universal. The question is, What allows us to make intelligent choices? In a complex world, how do we make sensible design decisions that positively shape a society?

GETTING TO DENMARK

Both of us—Jens Martin and Christian—have each worked extensively with and for the design field for over two decades. From starting design agencies, labs, and teams and crafting everything from award-winning bicycles to national policy, we have not only witnessed but also harnessed the power of this ever-emerging field:

INTRODUCTION

Jens Martin, mostly, in the fast-moving world of industrial and digital design; Christian, mostly, in the complex domain of policy innovation and governance. Both of us with a strong international engagement, from the TED community to the World Economic Forum and the European Union. Both of us as leaders, as design advocates, as speakers, authors, and educators. Each of us in our own ways have worked to expand how we think about design. But before writing this book, we had never put our minds together. Then one day, Jens Martin approached Christian and asked if we shouldn't write *the* book on Danish design. We sat down and talked. Then talked some more. And as we discussed and debated the topic, we recognized that the evolution of design in Denmark reflected a much bigger question: How might we *expand* design so that it remains an indispensable way to address the unprecedented levels of global turbulence and change we're all seeing?

That brings us to yet another reason for seeing design as a way to create a better world: we've seen what it has done to our own country. Denmark is routinely ranked as one of the world's happiest nations. Why? Because Denmark is a rich and healthy country in a region that has enjoyed decades of peace and prosperity. It enjoys comparatively little poverty and economic inequality and relatively high levels of gender equality and social mobility. And it has little corruption and a high degree of societal trust. Indeed, in Denmark, it's not uncommon to see parents leaving their babies asleep in strollers parked outside shops and cafés.

In his 2014 book *Political Origins and Political Decay*, the American political scientist Francis Fukuyama used the phrase "getting to Denmark" to describe how to make a country resemble this little nation of ours. The question for us is, How did *Denmark* "get to Denmark"? We maintain that much of the strength of the Danish model did not come about by chance. Nor is it the result of some special trace element in Danish water. In short, it's because Denmark was *designed* that way.

That's right. Regardless of what a hundred lifestyle publications might have you believe, it's not (just) because of *hygge*, the quintessentially Danish notion of coziness or conviviality. It's because Denmark is a thoroughly well-designed society. And we're not just talking about furniture and architecture—the things we most associate with Danish design—but industrial design, such as thermostats and wind turbines, and service design, like digital interfaces to make interaction with government bodies quicker and easier. Design underpins the transparency that lets Danes take care of everything online, from parking permits and medical appointments to driving licenses and marital separation. (Indeed, you can agree to get divorced and obtain your divorce papers digitally. So seamless is it that couples with children must now take a mandatory thirty-minute online course to stop and rethink whether they are ready to seal the deal.)

By taking a look at the well-designed operating system that Denmark runs on, we believe we can learn something universal about the power of design for shaping a society in meaningful and positive ways. Take, for instance, urban planning. Copenhagen was once a car-clogged city on the verge of bankruptcy. City planners saw an alternative future for the capital, one in which residents relied less on gas-guzzling cars and more on sustainable forms of transportation like walking and cycling. Planners pedestrianized much of the city center, added cycling infrastructure—including miles of safe, segregated bike lanes—and helped turn Copenhagen into one of the world's wealthiest and most livable cities.

What Denmark achieved can in theory be matched by other countries. (In fact, the notion of making a city more like the Danish capital—"getting to Copenhagen," as it were—even has its own term: "Copenhagenization.") Even so, this book isn't about Danish approaches to design. Rather, it is a book *informed* by the design culture and impact we see here. There are lessons to learn from Denmark, and we'll share examples that we know firsthand because this is, after all, where we live. But we'll provide many examples from beyond our borders too, not least because we believe the world can be improved by design solutions based on a wide variety of values, contexts, and perspectives—including, but not limited to, our own design heritage.

Denmark also shows us that we can change things if we want to—that we are masters of our own destiny. The anarchist David Graeber observed that "the ultimate, hidden truth of the world is that it is something we make, and could just as easily make it differently"—yet for more than a quarter of a century the world has been mesmerized by a monolithic and deeply ideological approach to innovation, technology, and design that all but denies our agency in making the world. We look at this pernicious ideology next—and explain how expansive thinking can lead the fight back.

AGAINST TECH DETERMINISM

In September 1995, an obscure British magazine called *Mute* published a paper about "an emerging global orthodoxy concerning the relation between society, technology and politics." The authors, Richard Barbrook and Andy Cameron, called this orthodoxy "the Californian ideology" after its state of origin. "By naturalising and giving a technological proof to a libertarian political philosophy, and therefore foreclosing on alternative futures," they wrote, "the Californian ideologues are able to assert that social and political debates about the future have now become meaningless."

Barbrook and Cameron were referring to Silicon Valley. The ideology—technological determinism—had already been "embraced by computer nerds, slacker students, 30-something capitalists, hip academics, futurist bureaucrats and even the President of the USA himself... With no obvious opponents, the global dominance of the Californian ideology appears to be complete."

Little has changed since. Describing technological determinism as "the polite term for the delusions that grip the lords of Silicon Valley (and their fans elsewhere)," the *Guardian*'s technology editor, John Naughton, has argued that the ideology may explain "why they have manifested such blithe indifference to the malign effects that their machines are having on society. After all, if technology is the remorseless bulldozer that flattens everything in its path, then why waste time and energy fretting about it or imagining that it might be controlled?"

Take, for instance, the ride-sharing firm Uber. It illegally operated autonomous vehicles (AV) in San Francisco—until one of them ran a red light. Uber moved its AV program to Arizona—yet in March 2018, one of its AVs killed a pedestrian. Uber simply put the program on hold for six months before relaunching it in Pittsburgh.

Lost amid the tragedy and the controversy is any critical analysis of the underlying technology and the need for it. Where is the discussion about the impact of driverless vehicles on our quality of life in cities? If anyone can simply jump in an autonomous vehicle, what might the consequences be? Gridlock? More street space taken up by cars and other driverless vehicles? Less mobility? Fewer cyclists and pedestrians? More urban sprawl, thanks to a reduction in the inconvenience of long commutes?

INTRODUCTION

Who knows? And yet politicians and pundits alike parrot the line that driverless vehicle technology is inevitable, that it necessarily represents progress. It isn't and it doesn't. Ditto with drone deliveries, lab-grown meat, and AI. As fascinating as all these technologies are, the way they're implemented often become captured by narrow and short-sighted interests.

Meanwhile, the tech giants spin a different narrative. "Sidewalk Toronto is about improving people's lives, not developing technology for technology's sake," wrote the CEOs of Waterfront Toronto and Sidewalk Labs, a subsidiary of Google's parent company, Alphabet, in the *Toronto Star* in 2017. They were announcing big plans for the Canadian city: to build a "digitally wired neighborhood of the future" on the edge of Lake Ontario. By combining data-gathering sensors with cutting-edge urban design, they would supposedly consign congestion, unaffordable housing, and excess emissions to the dustbin of history.

These are sweet-sounding words—but at what cost? Opponents of the plan, like Toronto-based privacy advocate Bianca Wylie, say data-driven approaches to community planning by private companies, not governments, are fundamentally wrong. "It's about our neighborhoods, our cities, how we want them to work, what problems should be solved, and which options should be looked at," Wylie told *CityLab*. "I reject the technocratic vision of problem solving."

Wylie's was a rare voice of concern. As the *Guardian* warned in January 2019, tech firms are consistent in pumping out a crude narrative about AI: "While there may be odd glitches and the occasional regrettable downside on the way to a glorious future, on balance AI will be good for humanity. Oh—and by the way—its progress is unstoppable, so don't worry your silly little heads fretting about it because we take ethics very seriously."

EXPAND

Joichi Ito, the former director of the MIT Media Lab, made a similar argument in "Resisting Reduction: A Manifesto," an online broadside against the cult of singularity—or the idea that "AI will supersede humans with its exponential growth, and that everything we have done until now and are currently doing is insignificant."

Ito argues that "the idea that we exist for the sake of progress, and that progress requires unconstrained and exponential growth, is the whip that lashes us," adding that "the Singularitarian view that with more computing and bio-hacking we will somehow solve all of the world's problems or that the Singularity will solve us seems hopelessly naive." Ito further argues that "we need to embrace the unknowability—the irreducibility—of the real world that artists, biologists and those who work in the messy world of liberal arts and humanities are familiar with."

We agree. While some technology is unquestionably life-changing, the belief that technology is what mainly drives history—and that the future is locked into a particular path—has captivated too many people for too long. We argue that it's time to resist the siren song of teleology, the belief that history has an inevitable direction.

As champions of design, nothing matters more to us. If necessity is the mother of invention, then determinism is its prophylactic. If we accept that things are going to keep moving in a specific direction, why put any more time or energy into changing them? Determinism removes human agency and politics from the equation and holds that there are no choices to be made. But as Jill Lepore put it in the *New Yorker*: "Machines aren't something that happen to us; they're something we make."

And yet, as the scope of our problems is expanding, with ever greater consequences for people and the planet, the scope

INTRODUCTION

of innovation has been getting narrower. In fact, it's worse than that. Innovation is getting faster, allowing those of us outside the monotechnological bubble of Silicon Valley little chance to keep up, to monitor, or to choose. Big Tech is often compared to nation-states in terms of the global power that the behemoths of Silicon Valley now wield. In 2017, Denmark made history when it appointed Casper Klynge, a long-serving diplomat, to become the world's first tech ambassador. (He's since been succeeded by a young woman, Anne Marie Engtoft Larsen.) Switzerland, Canada, Austria, and other countries are following suit. Governments see that Big Tech has country-like powers and are figuring out how to deal with their un-country-like structures. Diplomats could help. Ironically, it is countries that are providing diplomats to the tech world, and not vice versa. This says a lot about the democratic challenges we face.

Fortunately, there are pockets of resistance, voices in the wilderness such as Wylie's and Ito's. Consider, for instance, the Copenhagen Letter, a kind of manifesto published in 2017, following a tech festival in the Danish capital. Addressed to "everyone who shapes technology today," it states: "We live in a world where technology is consuming society, ethics, and our core existence. It is time to take responsibility for the world we are creating. Time to put humans before business. Time to replace the empty rhetoric of 'building a better world' with a commitment to real action."

The Copenhagen Letter has been signed by over five thousand technology leaders, designers, and decision makers—including us. One of its five principles is a call to move from human-centered design to *humanity*-centered design. We agree that it's time to take back control of our future—and that design holds the key. Too many people talk about technological developments as though they're inevitable. They're not. It's always a matter of choice. We can change the course of the future—but only if we want to.

It's worth recalling, though, that Silicon Valley hasn't always dominated the debate. In the twentieth century, there were competing visions of society, each with its own impact on the development of technology. The Soviet Union developed groundbreaking technology such as Sputnik satellites and turbojet trains, while Gaullist France built big centralized systems such as the TGV and Minitel (a kind of proto-internet). Their influence is long gone now (along with that of communism and French philosophy), but they remind us how novel thinking about the use of technology can serve society—and that different paths are available.

They also remind us that as we expand our thinking about designing solutions to modern problems, we might want to look backward as well as forward.

Phage therapy, for instance, provides a fascinating case for the merits of retrospection. It involves using bacteria-specific parasitic viruses known as bacteriophages to kill pathogens. Only recently has it been revived as a global alternative to deal with multidrug resistant infections.

Viruses that kill bacteria might sound like something out of the *War of the Worlds*. Bacteriophages were reported by some scientific communities in dysentery patients shortly before they began to recover. In the Soviet republic of Georgia, the microbiologist George Eliava spearheaded research and founded the Eliava Institute in Tbilisi to develop phage therapy. The approach became prevalent in the Soviet republics and is in widespread use in Russia, Georgia, and Poland to this day. In the West, however, an alternative way of treating bacterial infection, antibiotic penicillin, was being developed in parallel. And due to their success with antibiotics (and perhaps a rivalry with Soviet scientists),

INTRODUCTION

Western scientists largely gave up further use and study of phage therapy.

Secluded from Western advances in antibiotics, Soviet scientists continued to develop their already successful phage therapy and to improve treatment and research. Due to the scientific and cultural barriers of the Cold War, this knowledge was never translated or transferred and did not proliferate around the world. In fact, according to medical historian William Summer, in the postwar period, phage therapy acquired a "Soviet taint" and was deemed "scientifically unsound because it was politically unsound." Today, phages are increasingly recognized as a viable alternative to antibiotics—not least as a solution to the rapid rise of antibiotic resistance.

Although later turns of events—such as the end of the Cold War—are a positive development, there are still multiple examples of cultural and ideological biases that potentially prevent technological diversity today. The initial resistance of the Western scientific community to the Russian Sputnik V vector vaccine, also known as Gam-COVID-Vac, has been speculated to be in part geopolitical. Expanding our thinking just might get us beyond such barriers.

Yet for more than three decades, we have lived in a unipolar world dominated not only by American economic, political, and military power, but also by a largely US-centric vision of the role that technology can play in driving business and societal innovation. For a long period of time, we lived in a mono-technological world.

Today, the world is once again becoming multipolar, certainly in both the political and economic sense. Politically, alternative centers of power have emerged and asserted themselves, including the European Union, Russia, China, and India. Economically, the EU already represents a much larger market than the United States, and China will likely soon surpass the latter as the world's largest economy. And by 2040, the Middle Kingdom will represent around a quarter of the world's total economic output.

Along with this renewed political and economic multipolarity, there is a parallel shift in the forces of innovation. When it comes to visions for how emerging technologies can serve people, businesses, and society at large, there are more alternatives out there, including in Africa, Asia, Europe, and the Nordic countries.

For instance, some African countries are pioneering the use of drones to deliver humanitarian supplies to far-flung rural areas. In India, the notion of frugal innovation has led to radically more efficient models in health care, such as eye surgery. Japanese firms have paved the way in bringing augmented reality (AR) to computer gaming. China has stolen a march in the application of AI and digital facial-recognition technology. Iran has taken a lead in developing ground-effect aircraft and now uses this technology in its armed forces. And the Nordics are arguably front-runners when it comes to sustainable energy solutions. Seeing the technological advancements of each of these regions through a magnifying glass makes regional biases and values become apparent. The question arises: Does this technological diversity come from cultural diversity only, or can technological diversity lead to cultural diversity too?

Though all of these technologies derive from different societal priorities and cultures, they are themselves "ideology agnostic." Once they exist, they are available to us to apply in ways that could change the world for the better. Drawing inspiration from alternative models of society and innovation doesn't require us to endorse a particular political system or ideology. The point is that we ought to diversify technology, widen the bandwidth of innovative

INTRODUCTION

resources available to decision makers, and embrace a multipolar world once again. And to do so, we need to revisit one of the most influential business ideas in recent years: design thinking.

BEYOND DESIGN THINKING

For more than a decade, design thinking has been a hallmark of creative problem solving in a business context. However, the concept has increasingly been reduced to a simple set of methods and processes that anyone in principle can apply as a way to empathize with users, cocreate new ideas with others, and build prototypes of potential solutions. Design thinking has served to democratize the field of design in ways many professional designers had never imagined (and many are uncomfortable with). On a positive note, this has propelled an understanding of basic design concepts into C-suites around the world. On a negative one, it risks projecting a limiting and reductive image of what design not only is, but can be.

As we argue more thoroughly in the next chapter, the era of design thinking is behind us—and a new era of design is ahead of us. We believe that design, in its broadest meaning (which includes design thinking), has further to go if it is going to drive meaningful societal change, address our future challenges, and foster happy, healthy, sustainable, and prosperous lives.

We propose going further and maintain that we ought to challenge the idea of what it means to design—including where design activity can take place. That is, we believe we need to expand the field of design and stretch it even further across the realms of business, politics, society, and beyond—and that by thinking *differently*, especially about design—including *how* we design, *what* we design, *who* we design for, and *how long* we design for—we can develop tomorrow's truly innovative ideas and create a better world.

In other words, we believe that we need to go *beyond* not only design thinking (which has focused mainly on the power of design from a *management* perspective) but also human-centered design (which has focused overly on *method*) and engage in more *multidimensional* thinking. Indeed, what we propose is a kind of "design thinking on steroids"—or what we have chosen to call "expansive thinking."

Expansive thinking means imagining alternative futures and going beyond the safe, stale, and culturally determined mindsets that typically take root in existing systems, sectors, and organizations. It means innovating on a more *systemic* level, figuring out what people, communities, and ecosystems need as a whole, and testing, improving, and scaling new approaches. Expansive thinking means challenging assumptions and preventing intellectual inertia.

In short, we must not shy away from navigating complexity and ambiguity or from working creatively with knowledge gaps and the world's wicked problems to come up with new insights. To do so, we will need to advance from a mono-technological culture to become more technologically diverse, expand and transcend the current reality to explore unforeseen possibilities, going beyond what is already there or is thought to be predetermined. That is the essential transformation challenge that we face.

SIX EXPANSIONS

So, how can design be expanded to help us work more effectively on our most urgent problems?

INTRODUCTION

To get started, the next chapter defines design more thoroughly and dissects the current and present developments in design that this book takes as its jumping-off point. This chapter can be skipped by readers familiar with the design field and its current debates.

We then propose six expansions: six ways of thinking about the role of design and innovation; six ways to broaden your thinking about the topics and issues that influence your decisions and behaviors. Some are relatively mainstream ideas, such as shifting from the traditional "take, make, waste" model of production and consumption and embracing circular and networked models of value creation. Others may be less familiar, like expanding our thinking about designing life itself.

The first expansion concerns **time**. We live in an ever-faster world: a world of "flash crashes" and short-term results, fast fashion and limited shelf lives. Yet many of our biggest challenges require long-term thinking and planning. Whether it's shifting to renewable energy sources or creating a sustainable food system, our most pressing problems can't be fixed within a fiscal year or a four-year election cycle. They're problems with a ten-, fifty-, or hundred-year time frame. From hospitals built for the next century of health care to clocks designed to run for ten thousand years, the chapter showcases new ways to expand our horizons and design for the long-term—not the next season.

The second expansion concerns **proximity** and how we might design for what's perceived as close and for what's considered far away. We look at ways in which artists and designers have sought to compress time and space, from blocks of glacial ice dumped in capital cities to exhibitions that put visitors in the shoes of refugees. We make the case for designing for people and communities at a hyperlocal level—to make concrete, place-based interventions that take account of how people make sense of their lives in very particular contexts—and also to see design as a way to deliver change on a wider scale in response to global problems such as poverty, climate change, and urbanization.

The third expansion concerns **life**. It asks us to expand our thinking about life beyond the limits of what we've long thought possible, and to rethink the boundaries between life and death, digital and biological, as we increasingly contemplate a world of cloning, prototyped body parts, and the extension of existence in the digital afterlife. We also look at new ways of designing with living materials, such as bacteria and algae, as well as at the inevitable expansion of what we understand to be "living," in the form of synthetic meat and robots. Finally, we propose expanding the scope of design from humankind to the planet.

The fourth expansion concerns **value** and how we extract, trade, own, consume, use, and discard resources. From cigarette butts in Taiwan to beer bottles in Copenhagen, we showcase mind-bending examples of finding the value of everything and make the case for expanding our thinking in three ways: first, by introducing more dimensions of value into the arena of production; second, by shifting from linear to circular models of business and institutional design; third, by seeing value creation not as a chain or even a circle but as networked systems where multiple actors interact in collaborative models of value creation.

The fifth expansion concerns what one could describe as **dimensions**. Encompassing everything from nanotechnology and wearables to AR/VR and continent-wide energy grids, this chapter explains how questions of design are no longer concerned only with the physical representations of ideas, but now straddle both the physical and digital worlds—and increasingly the latter. Indeed, more designers are manipulating bits and bytes, not just

INTRODUCTION

atoms and molecules, and are designing artificial intelligence and digital platforms. Just as the idea of universal design became crucial to the physical world, so do we need a new understanding and approach to help us design in the immaterial world.

The sixth expansion concerns **sectors** and our understanding of the roles of the public and private sector in design. The traditional domains we use to categorize social and economic activity—public, private, civic—are blurring. An expansion of the roles, contributions, and interactions between these sectors is increasingly crucial to designing for sustainable change. In this chapter, we make the case for seeing the public and private sectors as complementary sets of resources, each with its own opportunities and constraints. We also argue that we need to let go of industrial society's notions of industries and sectors—and instead think much wider in terms of ecosystems and value-creating relations among actors.

The book ends with some reflections on the implications of these expansions for the future of design and how we might go about the kind of deep transformations that businesses, societies, and individuals need. It includes three hypothetical consequences of applying expansive thinking to the future of products, flight, and urban mobility. The appendix contains a series of questions pertaining to each of the six chapters, which are designed to kick-start your own reflections about how to apply expansive thinking.

Expansive thinking is therefore about taking our usual ways of looking at the world and seeing what would happen if we pushed them much further. In that respect, this is not so much a book about design as one about *how to think*. As such, it is intended for anyone who acknowledges that the world is in dire straits and that we face many complex problems. It is also a book for people who realize the dangers of leaving innovation in the hands of Silicon Valley only; a book for anyone worried about the next twenty years—and curious about how to change it. It is a book for anyone hoping to make a difference and transforming society, as well as for anyone seeking to cope with political and economic volatility. In short, it is a book for change makers—artists and architects, engineers and designers, entrepreneurs and investors, leaders and executives—and for all those who wish to become the change makers of the future.

Faced with headline after headline about dysfunctional services, the Department of Veterans Affairs adopted a veteran-centered design approach. In time, it improved its benefits-claims process, modernized its call centers, and upgraded the forms that veterans use to contact mental health services. The VA also rolled out an agency-wide framework to more accurately gauge its users' needs. It has since seen a 25 percent increase in trust among veterans. (In a 2017 interview with the *New York Times*, Brooks—who is now a design executive and distinguished designer at IBM—described trust as the department's "share price.")

The approach we advocate in this book—expansive thinking—is nothing less than an attempt to supersize that kind of design. We believe that with its broader perspective, only expansive thinking can unleash the truly innovative ideas of tomorrow and create a better and more prosperous future. If the design-led changes at the VA showed us what's possible, expansive thinking will help us accomplish things we didn't even realize were possible.

Speaking to the *Guardian* in 2016, Nicholas Stern, the former chief economist at the World Bank and chair of the London

INTRODUCTION

School of Economics' research institute on climate change and the environment, warned that "people have not sufficiently understood the importance of the next 20 years. They are going to be the most decisive two decades in human history." We've already lost a quarter of that time—so let's get cracking.

1
DESIGN PAST AND PRESENT

Design is too important to be left to designers.
Raymond Loewy

Design won't save the world. Go volunteer in a soup kitchen, you pretentious f**k.
Saul Griffith

This book seeks to expand how we think and act as we create humanity's responses to the grand, complex transitions the world goes through right now. In the introduction, we argued that design is the jumping-off point toward the expansions that will help us solve problems differently. So to get there, we first need to consider what design is, what it can do, and what kind of problems it can solve. (If you are familiar with the topic, skip this chapter.)

EXPAND

WHY DESIGN

Let's begin with a basic definition. Design is an intentional discipline that uses form, sense, and sensibility to connect people's needs with what is technologically feasible and commercially or socially viable. As such, it affects contemporary life in at least four areas: symbolic and visual design; the design of material objects; the design of activities and organized services; and the design of complex systems or environments. Design cuts across all other human activities as a particular concept that addresses how physical, commercial, social, and public outcomes are created.

Today, in fact, design signifies an ever-widening set of approaches that can be deployed at vastly different scales to focus on a broad range of problems. Just look around you. Design covers everything from artifacts, services, and platforms to systems, policies, and models of business and governance. The internet is designed. Google and Facebook are designed, from the user interfaces we experience to the algorithms, systems, and business models underpinning them.

Of course, we haven't always seen design like that. The role of design in society has come a long way fast. Historically, design was linked to industrialization and the rise of mass production. Designers created things that reflected our tastes and demands. In fact, for most of the past century, the profession was largely about the creation of graphical communication and physical objects—from branding and marketing to industrial design and consumer products. For instance, craftspeople made chairs, first according to their own tastes, and later according to the specs sent over by a designer, as design became a distinct function in support of industrial manufacturing.

In the past few decades, however, design has splintered. First, the profession moved beyond the creation of physical products and graphics to encompass the realm of services and systems. Instead of striving to make beautiful things, the designer began to help organizations think about their products and services deliveries, for instance, through tools of visualization that exposed the interactions between the organization and its external users. And as we shall see shortly, managers looking to do things differently started to apply the principles of design thinking.

Design has also become more multifaceted, spawning exotic-sounding subdomains such as interaction design, critical design, environmental design, and biodesign—often with huge global impact. User-experience design, for instance, has become a driving force in Silicon Valley's tech economy as well as in European states promoting the so-called sharing economy.

As for *who* is a designer, well, we'd go as far as to say that we all are—potentially. "Everyone designs who devises courses of action aimed at changing existing situations into preferred ones," wrote the American Nobel-winning economist and decision scientist Herbert A. Simon. Chances are that if you're reading this book, you're a designer in that most expansive sense of the word—or at least you can be, if you wish.

EXPANDING DESIGN THINKING

The most obvious consequence of the splintering of design is that it has opened up the field to much wider areas of organizational impact and value creation. In fact, design is increasingly seen as a way to devise strategy and achieve change at all scales and levels. And there is perhaps no better indicator of its expansion than the growing popularity of design thinking in corporate circles since the early 2000s.

Simply put, design thinking is about applying the principles of design—like prototyping and empathizing with the end user—to the way that people work. Typically, it uses physical artifacts to explore, define, and communicate the challenges at hand. For example, the Department of Veterans Affairs used a set of "customer journey maps" to understand how veterans interacted with it.

Design thinking is part of the A/D/A (architecture/design/anthropology) model. It is a school of thought that focuses on a collaborative and iterative style of work and a synthetic mode of thinking, compared with the practices associated with the traditional, analytical M/E/P (mathematics/economics/psychology) management model. A basic outline of the problem-solving process of design thinking goes like this: identify, interpret, investigate, ideate, illustrate, imply, iterate, implement.

One telltale sign that design thinking is taking place nearby is an orchard of Post-it notes on the wall. As *Fortune* put it: "The stickies are a hallmark of 'design thinking' exercises, in which participants often jot down thoughts on the brightly colored pieces of paper and place them on a whiteboard as part of creating, for example, an 'empathy map' to understand the perspective of the user or customer by imagining what she or he thinks, feels, says, and does."

Described by MIT's Sloan School of Management as "an innovative problem-solving process" with several steps—from understanding the problem, to exploring the possible solutions, to prototyping, testing, and iterating—design thinking exploded in popularity in 2008 when Tim Brown, CEO of design company IDEO, wrote about it in *Harvard Business Review*. It soon became the darling of the C-suite, adopted with great success by the likes of Procter & Gamble and General Electric.

As leading design advocate Bruce Nussbaum wrote in *Fast Company* three years later, design thinking "originally offered the world of big business . . . a whole new process that promised to deliver creativity." And he extolled its power to "move designers and the power of design from a focus on artefact and aesthetics within a narrow consumerist marketplace to the much wider social space of systems and society." Similarly, design expert Helen Walters—who's now head of curation at TED—described design thinking as providing "insights galore," adding that "there is real value and skill to be had from synthesizing the messy, chaotic, confusing and often contradictory intellect of experts gathered from different fields to tackle a particularly thorny problem."

But here's the thing. The world's problems are getting thornier by the day—and design thinking is no longer sufficiently fit for purpose. From climate change and emerging technology to population growth and rapid urbanization, we are living in an age of unprecedented complexity and change, turbulence and speed. And in order to tackle the thorniest problems, we need tools that are more powerful and more creative than design thinking alone.

Design thinking has several shortcomings, including its lack of built-in criticism and its neglect of any consideration of aesthetics. But perhaps its biggest problem is its consideration of time horizon—or lack thereof. We now live in a world where development in most sectors is accelerating quickly. Products that once took months, if not years, to develop are now being prototyped, validated, adapted, and brought to market faster than ever—sometimes within a few weeks. At the same time, we face major long-term challenges, such as climate change, preventing the next pandemic, and seismic shifts in technology and politics.

All of which means that we need to bridge ultra-short-term market needs with long-term challenges, some of which may yet stretch over decades. And yet the design-thinking method really only responds to medium-term challenges. It is incapable of handling either very long-term or very short-term issues—and certainly not both at once.

A lack of speed is inherent in design thinking. The process relies on anthropological research with human beings as a benchmark—an approach that is often long and cumbersome. The usual customer journey that clients are taken through can be time-consuming too, as can the process of delivering insights. And while design thinking's defenders might point to design sprints—a fast process that sets goals, tests, and assumptions with users, and picks a preliminary roadmap—as the answer, it remains the case that design thinking wasn't built to deal with large amounts of real-time data.

As businesses increasingly combine a rush to market with a desire to have a positive impact, design thinking looks increasingly out of touch with reality. Consider a company that makes headphones and wants to launch a brand-new product in two weeks' time. How would you use design thinking in that scenario? You would have to generate the requisite knowledge, cooperation, and prototypes within an oppressively short space of time—to say nothing of considering the longer-term impact of your decisions on people, the planet, and the company itself.

Last but not least, prototypical design thinking has a user-centric focus. As the turnaround at the Veterans Affairs department showed, user-centered design can be lifesaving. But user-centered design is no panacea either. An increasingly common critique of design thinking is that it often only considers "typical" or "average" user segments. But what about the "atypical" users that live at the edges? What about including the marginalized lives and voices? Further, as Mike Monteiro argues in his book *Ruined by Design*, what if a long-term challenge is in fact caused by the user? What if our focus on user-centered design—which is really to say human-centered design—comes at a cost to other forms of life on this planet we share? In other words, we should no longer assume that the user should be at the center of things. The age of orthodox user-centered design is coming to an end; as we explore later, the age of life-centered design has begun.

To be sure, we aren't alone in seeing design thinking's shortcomings. In 2011, Nussbaum called time on "the decade of design thinking," rejecting the framework in favor of "creative quotient." Walters argued that same year that "design thinking won't save you." In 2018, in a 99U talk called "Design Thinking Is Bullsh*t," Pentagram partner Natasha Jen poked fun at design thinking's apparent obsession with 3M Post-its. Among other points, she argued that lack of critique in design thinking is highly problematic for achieving quality in design work. Likewise, Professor Roberto Verganti, in his 2017 book *Overcrowded*, suggests that the methodologies in traditional user-centered design fail to generate meaningful solutions that people truly love. And in 2019, John Maeda, the head of inclusion and computational design at Automattic—and a longtime champion of elevating the role of design within organizations—raised eyebrows when he declared that "in reality, design is not that important" and relegated designers to a supporting role.

In a sense, the next evolution of design involves thinking differently about thinking itself. Adding the label "thinking" to design has been instrumental in propelling the discipline into all kinds of realms and organizations. We believe this is just the first step in unfolding the power of design. Finding ways to describe the *thinking* involved in design could help us expand the boundaries of design and reshape our future.

On one level, this means that instead of further developing the approaches and methodologies in design, we should *think about the thinking* that goes into design—in other words, about the *philosophy* of design. What do we mean by that? Well, design always involves some of the following factors: (1) form, sense, and sensibility; (2) functional and technological feasibility; and (3) commercial or social viability. A design philosophy should cover how design relates to issues such as perception, function, intentionality, recognition, game theory, narration, path dependency, innovation, nature, and economy. Interestingly, philosophers have covered all of these fields extensively, one by one, over the millennia, so the academic groundwork is there to help us expand our thinking.

At the same time, thinking about design thinking will require us to understand people and humanity, other species, and life itself. That is, we need to abstract from human-sized stories and consider what makes something meaningful to humans *and* other life forms. For instance, how do we avoid measuring everything in human lifetimes, human viscosity, or human-sized objects? To create a more just, sustainable, and prosperous world, design thinking must evolve with it.

2
EXPANSION 1—TIME

> We need a dream-world in order to discover the features
> of the real world we think we inhabit.
> **Paul K. Feyerabend**

> Civilization is revving itself into a pathologically short attention span.
> **Stewart Brand**

Founded in 1379, around the time the outlaw Robin Hood first appeared in myth, the New College is one of Oxford University's oldest colleges. Inevitably, it boasts legends of its own—including the story of the oak beams that span the ceiling of the college's ancient dining hall. The tale begins about a hundred years ago, when an entomologist discovered that the beams were riddled with beetles and needed replacing.

The fellows of New College were stumped. How on earth would they afford the timber required to make comparable beams? When one of the fellows proposed using the oak trees growing on the college's land, the college forester replied: "Well, sirs, we was wonderin' when you'd be askin'." The fourteenth-century fellows of New College had known the beams in the dining hall would one day need to be replaced—and so they had planted a grove of oak trees precisely for that purpose. The fellows' plan required expansive thinking about time—planting seeds for a result more than five hundred years in the future.

Thinking expansively about time means resisting short-term thinking and instead reconsidering the time frame for which we plan for the future and make decisions. As the tall tale of the New College oaks suggests, our first expansion concerns time. It makes the case for resisting short-term thinking and instead broadening our horizons in terms of the time frame for which we plan for the future and make decisions. In particular, we ask how far into the future we might learn to think. And we examine time-honored methods for expanding our imaginative faculties that are rarely used in decision-making processes today. They include taking inspiration from science fiction, Bruce Sterling's notion of "design fiction," and the use of scenarios to imagine alternative futures. But first, we venture into our contemporary world of design and discuss its growing obsession with short-term thinking.

EXPANDING BY SLOWING DOWN

When we asked David Fellah, founding partner of Manyone, a leading international design agency, about the acceleration of the product development cycle, he sighed. At the turn of the century, his previous agency Designit spent three years designing a phone for Bang & Olufsen, the Danish high-end audiovisual manufacturer. A decade or so later, it worked on a phone for Samsung—with a development time of just two months. Digital development,

Fellah adds, is no different. "There's a requirement to deliver a digital service within six to nine months where it used to be eighteen months, and soon it will be three months," he says. "There's a need for speed."

You see, instead of designing for the future, we increasingly design for the here and now. Why? Because businesses feel increasing pressure to cater to ever-shifting customer needs, which in turn puts more pressure on product development and truncates traditional design phases such as ideation and research. Increasingly, too, designers skip steps and traditional processes like applying their intuitive insights to iterative prototyping. Sometimes the iterations become the implementation, and what used to be beta phases are accepted as end results. This leaves the world with lower-quality products and flawed digital services.

Moreover, emerging technology, such as machine learning, is playing a growing role in this shift. Increasingly, for instance, companies put prototypical (or even hypothetical) products to market—"selling" them online and seeing which ones do well. The successful products are put into production. The others aren't, and the customers who bought them are refunded. According to this paradigm, design companies are anchored around rapid, extreme prototyping, machine learning, and imaginary outlets.

Consider what's happening in China, where e-commerce giant Alibaba is effectively operating "like the world's biggest focus group." That's because it can test products in consumers' newsfeeds and search results, based on their profiles and real-time purchase behavior. Leveraging Alibaba's platform is how Mars worked out that people would indeed buy a spicy Snickers bar—and how Unilever knew to launch a new line of deep-cleansing cosmetics for urbanites. Unilever's then-CEO Paul Polman told audiences in Davos in 2019 that Alibaba's data researchers had found that residents of Chinese cities were increasingly interested in pollution-fighting personal care products—a market with few midpriced options. Acting on this, Unilever came up with forty-eight prototypes of the antipollution skin cleansers at different price ranges—and showed them on Alibaba's online shopping portals *as if* they had been developed. Based on the actual purchasing decisions of tens of thousands of consumers, Unilever whittled down its prototypes and seamlessly launched a line of price-conscious antipollution products, starting with a cleanser.

Despite the strong case for user-centered, collaborative, and participatory design—now more than ever—technological developments are fast compressing the design process. Participatory design used to mean taking care to involve a broad range of stakeholders, including end users, of course, in a collaborative process. Emerging technology is increasingly driving the process and accelerating the time to market. This pressures design methodologies to reinvent themselves as shorter and shorter deadlines challenge designers to perform the steps typically involved in user-centered design in shorter time.

Meanwhile, our linguistic and imaginative capacities appear to be letting us down when they are most needed. In an age of fast design and even faster fashion, we are losing the ability to articulate the longer term. This includes long-term changes that may end up destroying us: "People are struggling to find language in the climate debate," the Danish-Icelandic artist Olafur Eliasson told us. "For instance, how do we make ten, twenty, thirty, forty, or fifty years palpable?"

For Jeff Bezos, the answer lies in horology. Amazon's founder is behind plans to build a clock deep inside a Texas mountain—a clock that will remain intact, operational, and accurate over ten thousand years. The clock is intended not for ramblers who have lost track of time but to encourage us to measure the long term in centuries—so that we might take long-term responsibility. "I believe long-term thinking is an extraordinary lever that lets people accomplish things they couldn't do otherwise," Bezos told *Wired*. Global poverty might seem impossible to solve in the next five years. But how about in the next hundred? Given a much longer time frame, might we not think differently about the problem and try out different solutions?

The ten-thousand-year clock is the brainchild of inventor Danny Hillis, who cofounded the Long Now Foundation to promote long-term thinking and "provide a counterpoint to today's accelerating culture and help make long-term thinking more common." In a similarly imaginative vein, Beatrice Pembroke and Ella Saltmarshe launched the Long Time Project in 2018 "to stretch our capacity to care about the long term," arguing that "the tunnel vision of short-term thinking" is "leading to decisions that might mean we are only left with a short term as a species" and that "developing longer perspectives on our existence will change the way we behave in the short term."

The question is, *How* do we learn to expand our thinking, start working on radically longer timelines, and designing for the long term? We believe that several intellectual tools can help, including science fiction narratives and "design fiction." But we begin with the use of scenarios.

EXPANDING THROUGH SCENARIOS

Since the 1970s, think tanks like the California-based RAND Corporation and its spin-off, the Institute for the Future, as well as companies such as Shell have used so-called scenario methodologies to improve their long-term strategic planning. Governments have embraced scenarios too. Singapore has scenario planning embedded in its Civil Service College, while the Canadian federal government has Horizons Canada, a twenty-strong team that uses foresight to "explore plausible, alternative futures and identify the challenges and opportunities that may emerge."

In many cases, the systematic use of scenarios has enhanced long-term thinking and consideration of possible consequences of choices in both policy making and corporate circles. Yet scenario work is often a highly analytical exercise. Roughly speaking, it entails mapping hundreds of emerging trends across social, technological, economic, environmental, and political factors. Experts are often deeply involved in the process—through, for instance, Delphi surveys, workshops, and seminars.

Scenarios are typically created by identifying and deciding which factors are most important for a future thematic such as "the future of health" or "the future of the chemical manufacturing industry." The timeline for a scenario typically ranges from a couple of years to several decades or longer. For instance, a well-known Danish foundation working in the pharmaceutical field recently created hundred-year scenarios. A key premise in scenario work is that the involvement in creating scenarios is in itself an eye-opening exercise. To the extent that key decision makers participate in scenario-building workshops, they are more likely to be influenced by them and take up the insights elicited by the alternative scenarios.

However, the scenario method can also be disappointing. The brainy, analytical approach can feel abstract, technical, and stale, and the resulting products (which are mainly written reports) can fail to engage a wider audience—especially if the hope is that the scenario will trigger immediate action and change. One of this book's authors (Christian) once advised an internal government scenario team and asked its manager what kind of impact their work had across the relevant parts of government.

"Hardly any," he said. "People might at best read our reports, but then nothing happens." That response is especially disappointing given the significant amount of time and energy usually spent on solid scenario development.

However, in recent years the analytical rigor of mapping trends and uncertainties—which is at the core of foresight work—has increasingly been supplemented with other approaches, not least from the design field. Bringing design—which is to say the ability to visualize and shape plausible futures in ways we can see and engage with—together with foresight methodology addresses some of the weaknesses of the method. By using collaborative design approaches, a wider group of people with more diverse professional backgrounds and experience can be included in the scenario-building process, stimulating a divergence of ideas and challenging conventional thinking. This could include a range of "unusual suspects" like artists, philosophers, and clerics—or even customers and citizens.

Creating tangible visual and bodily "future experiences" also increases engagement with the issues at stake and pushes participating stakeholders out of their comfort zone through the visceral and emotional connection with the future.

Take the "Boxing Future Health" workshop held in Copenhagen in 2017. Participants were invited to lie on a bed in a small, cylindrical room, surrounded by pink pillows as soft as marshmallows. The room smelled of freshly cut flowers, the lighting was soft, and a pleasant and reassuring voice revealed that it was 2050, and participants were being treated by the recently launched Ministry of Root Causes, whose purpose was to eliminate risk factors for illness, promote healthy living, and prevent people from ever getting sick.

This scenario was one of four "alternative future experiences" facilitated by the Danish Design Centre during the workshop, which drew participants from a range of fields, including business, government, academia, and health care. Along with medical-device manufacturers, there were also representatives of schools of medicine and nursing. As part of the workshop, the design team created four experiential spaces that participants could explore. Each space used light, sound, props, and stories to depict a hypothetical scenario.

The point was to facilitate long-term thinking about health care. Several questions were on the table. How will we train the doctors and nurses of tomorrow? What products and services will hospitals need? What is the future of nursing and health care if people are increasingly going to be treated at home? How might AI and precision drugs change how we treat patients? What will "health care" mean in 2050? At an even higher level, the aim was to explore how one might prepare all stakeholders across the health system for a future in which the only certainty is that the skills they possess today won't necessarily be the ones they need tomorrow.

In a way, the Boxing Future Health project typifies the argument of this chapter. First, it's an example of how designers can devise a set of approaches that stimulate our imaginations, and expand the range of decision-making options available to us. Second, it encapsulates the kind of long-term thinking that matters more than ever today. That's because we're living in a time of immense change and complexity but are driving faster and faster—and paying less attention to the road ahead.

EXPANSION 1—TIME

In fact, after experiencing one of the four scenarios, one workshop participant said that she became "severely depressed" at the prospect of that particular future becoming reality. Hopefully, her motivation to prevent this scenario increased as a consequence.

Design approaches can also be used to run collaborative workshops and seminars that bring possible futures in play with many different kinds of actors and stakeholders, allowing more creative and ambitious ideas for new strategies, services, and products to emerge. Consider, for example, the workshops that culminated in the blueprints for Denmark's new children's hospital in 2016.

In 2011, the Ole Kirk Foundation—which was established in memory of LEGO founder Ole Kirk Kristiansen—agreed to help fund a new hospital for children. Built in the 1970s, Denmark's national hospital was getting increasingly crowded. Its children's wards were scattered across six buildings—leading one exasperated parent to complain that "when they move you from one ward to another, it's like arriving in another country."

"The physical layout was terrible," Anne-Mette Termansen, a management consultant with years of experience in the health-care sector, told us. In 2014, she was tasked with finding out whether it would be possible to build the world's best hospital for children, adolescents, and families—and what it would entail. Or, as she puts it, she set out to explore "how to design the future for sick children."

In particular, Termansen was responsible for designing "the future user experience, the physical design, and the overall transformation of the organization." Along with Bent Ottesen, a professor of obstetrics and gynecology, she started running a series of workshops to explore what the future children's hospital might look like—and invited patients, families, and hospital staff, including doctors and nurses.

At the first workshop, held in January 2015, the participants used colorful Duplo blocks to build prototypes of the new hospital. A workshop in October 2016 saw the involvement of representatives of the entertainment industry, including LEGO and Tivoli Gardens, as the focus shifted to user experience and user design, and getting "the first 10 minutes" at the hospital right. The following year, the participants met to discuss six proposed buildings. As one participant put it: "One incredibly important aspect of a camp like this is to define the building's main concept, because that's one of the things you can't change later."

These workshops often led to eye-opening and counterintuitive revelations. When one group proposed making hospital suites "just like home," with an open-plan kitchen and dining room for the patient's family to spend time in, there was widespread approval—until a twelve-year-old boy with cancer quietly pointed out that it would mean that he could not be around his family because the slightest smell of food would make him sick.

"All these doctors and nurses working with children for twenty years or more had not thought about it," Termansen recalls. (Another interesting revelation: instead of queuing for a buffet, patients would prefer smaller, tastier meals that look like those you get on planes.)

Underlying all of these workshops were several truths. The first was that while they were attempting to design the future, by definition, the future is unknowable. To put it another way, one of the only things they knew about it was that they did not know much about it. What the new children's hospital and its patients would need when it opens in 2024 is likely to be very different to what we might predict they'll need today, let alone in 2044.

Of course, several forecasts could be fed into the scenario. Demographic projections indicated that the new children's hospital would need to treat 17 percent more patients in 2025 than it did in 2014. Moreover, in 2016, Denmark raised the pediatric age limit from 15 to 18, meaning that a larger group of patients would be classified as children and be moved from adult wards to the children's hospital.

The workshop participants also considered the likelihood that it will become increasingly easier to treat patients at home. They were mindful, too, that the new hospital should be built with the greatest possible flexibility so that it could be adapted to unknown future needs. That meant future-proofing the hospital so that it could implement the highest technological standards. Among the options included in the blueprints: coordinated patient and medicine scanning, electronic medical records, electronic wayfinding, and continual monitoring of individual bed locations.

Denmark's hospital of the future will seek to inspire the private sector in several ways. First, an entire floor will house an innovation lab where medical students, researchers, and employees of medical or technology firms can work together and cooperate on designing new products, devices, or services. The hospital hopes to drive private innovation in another way.

"The medical industry is characterized by very deep specialization," Termansen says. "We want to build an operating theater that incorporates all of the specialisms."

For example, one company currently produces everything related to eye surgery, while another produces the entire suite for heart surgery. The new hospital hopes to change that. Likewise, it would prefer to have one microscope that can be used for surgery on, say, eyes and ears, as well as plastic surgery.

"That isn't yet on the market," Termansen explains, "but it's something we would love private companies to help develop."

Denmark's future children's hospital will be physically flexible too.

"We don't know how many operating theaters we will need in twenty years, or how many beds, outpatient rooms, so we can change among these capacities," says Termansen. The new rooms will all be the same size, allowing future administrators to switch between recovery rooms, outpatient rooms, or offices, as and when necessary. "We wanted it to be flexible so you could tear down walls or build them up, so mistakes can be corrected," adds Termansen.

All of these emerged from the workshops that Termansen ran. They are a testament to the power of user involvement and long-term thinking. As we will explore in the following section, the intersection of science fiction and design adds new dimensions to our imagination of potential futures.

EXPANDING THROUGH SCIENCE FICTION

Literature is well-suited for long-term perspectives, past and future. And science fiction, in particular, has a way of sparking the imagination and expanding our horizons. In fact, some believe it ranks alongside scenarios and design fiction as an indispensable part of any innovator's toolbox in the early twenty-first century.

Science fiction writers have long projected the possibilities of technological advances into the future and stimulated our notion of what the future might bring. For one thing, many technological achievements were foreseen by writers of speculative fiction: Mary Shelley's *Frankenstein* predicted modern transplants, Jules

EXPANSION 1—TIME

Verne's *Twenty Thousand Leagues Under the Sea* predicted the electric submarine and *Paris in the 20th Century* the Hyperloop, Aldous Huxley's *Brave New World* mood-altering pills, and Douglas Adams's *The Hitchhiker's Guide to the Galaxy* audio-translating apps. Meanwhile, the 1960s cartoon series *The Jetsons* predicted both the smart watch and video calls, while the Steven Spielberg movie *Minority Report* predicted gesture-based user interfaces, personalized advertisements, and predictive analytics.

Often, science fiction presages science fact, such as Arthur C. Clarke's proposal of using space satellites for global communications or the long-running television series *Star Trek* providing Motorola with inspiration for the StarTAC, which closely resembles the communicator used by the galaxy-exploring crew.

However, it would be a mistake to see science fiction as solely about forecasting the future. "No serious futurist deals in prediction," wrote Alvin Toffler in the introduction to his global best-seller *Future Shock*. "These are left for television oracles and newspaper astrologers." Today, what makes science fiction relevant to designers and decision makers alike is its ability to help us make previously unimaginable leaps.

Science fiction can also "provide quite startling insights into technological trends, and their potential impacts on society, when they are combined with real-world advances in science and technology." So writes Andrew Maynard in *Films from the Future: The Technology and Morality of Sci-Fi Movies*. For example, he describes *Jurassic Park* as "a wonderful jumping-off point for exploring modern-day genetic engineering," *Never Let Me Go* as a "chilling reminder of how easily a deeply-desirable technology can erode the moral soul of a society," and *Minority Report* as a morality tale about crime prediction just as "functional MRI scans, artificial intelligence and data analytics are all being used to try and pre-empt crimes before they happen." Maynard in particular emphasizes how science fiction films can teach us lessons for more responsible innovation—a growing field of research that seeks to assess the moral, ethical, and political implications of technology.

Today, a slew of companies are using science fiction to explore new technologies and their potential consequences—and to envision alternative futures in a fast-changing world. Other firms including Magna, a mobility technology company, and Perception, a cutting-edge motion graphics studio, explicitly talk about using "science fiction thinking" to envision new paradigms, possibilities, and strategies for their clients in, respectively, the auto industry and cinema.

Then there's the German tech giant SAP. In January 2019, the then-head of its innovation platform Next-Gen, Ann Rosenberg, made the case for using science fiction thinking and applying it to real business problems. In *Science Fiction: A Starship for Enterprise Innovation*, Rosenberg argues that it is "rapidly becoming an essential ingredient in the innovation process." Even *Harvard Business Review* appears to be on board. In 2017, it ran a piece titled "Why Business Leaders Need to Read More Science Fiction."

With its aliens and robots, spaceships and time travel, science fiction is all too easily dismissed as the stuff of fantasy. Yet science fiction thinking can offer a timely antidote to the technological determinism emerging from Silicon Valley, which we explored in the introduction. The orthodoxy championed by the tech giants of the Bay Area is that technology will save the day, if not the world; that emerging technologies are superior to preexisting ones; that progress is predetermined by various laws (like Moore's law, or the law of accelerating returns); and that the ethical consequences of, say, AI, which have not yet been fully thought through, will be answered by the invisible hand of capitalism. But too few tech designers consider the implications of their creations. Did the creators of Amazon's digital voice assistant Alexa think about its role in the emergence of a surveillance society? Or did they simply want to make shopping easier? The debates engendered in science fiction could better inform today's tech designers.

EXPANDING THROUGH DESIGN FICTION

Science fiction can help place the complex dilemmas of our age to the forefront and help us make more reflected choices in the present. It can provoke and expand our thinking by putting us into wild, imaginary future worlds. Where the field of science fiction becomes more limiting, however, is when organizations wish to become more concrete and build new solutions. When it comes to not only freely telling stories about the future, but also actually designing the prototypes that imply a changed world, something more is needed. That's where the emerging field of "design fiction" brings it all together. The term was coined by Bruce Sterling and then developed by Julian Bleecker, a researcher and product designer-engineer with Near Future Laboratory. Others have built on his premise that design fiction doesn't tell stories.

Instead, it builds prototypes that imply a changed world. In a seminal piece in *Wired* magazine—with the intriguing title "Patently Untrue: Fleshy Defibrillators and Synchronised Baseballs Are Changing the Future," Bruce Sterling, a leading figure in the cyberpunk movement, argues that it is design fiction—not science fiction—that can truly change the world: "[W]hen science fiction thinking opens itself to design thinking, larger problems appear. These have to do with speculative culture generally, the way that our society imagines itself through its forward-looking disciplines."

This book essentially has the same concern. We argue that the future is not fixed and cannot be predicted through our past experience—it is not predetermined by any law. We also argue that in order to shape the future we want, we need to expand our perspective—including on time—to be able to design more relevant solutions. Design fiction makes the same proposition. Or, in Sterling's words, because a design fiction is not real but can look and feel like it, it opens up alternative yet realistic opportunities for change. Sterling's argument goes as follows:

> *The objects offered to us in a capitalist marketplace have three basic qualities: they are buildable, profitable and desirable. They have to be physically feasible, something that functions and works. They need some business model that allows economic transactions. And they have to provoke someone's consumer desire. Outside of these strict requirements is a much larger space of potential objects. And those three basic limits all change with time. Through new technology, new things become buildable. Business models collapse or emerge from disruption. People are very fickle. That's how it works out—and the supposed distinction between "real" and "not-real" is pretty small.*

To expand our thinking, it is worth making the effort to not only build future worlds but to also populate them with structures and things that make them come to life, and that *might* exist someday. Having designed them makes it all the more likely they will become real. Thus, just like scenarios, design fiction projects us into the longer term to expand our perspective.

EXPANSION 1—TIME

If you'll forgive the pun, the long and the short of it is that it makes increasing sense to take the long view. And in theory, any entity could start doing so today. In the corporate world, Arup, Hershey's, and Volkswagen have all hired futurists to imagine the world we will inhabit in twenty, thirty, even fifty years' time. SoftBank Group, a multinational conglomerate founded in Japan in 1981, and one of the world's most valuable companies, has gone further still. In 2010, the company's billionaire founder and CEO, Masayoshi Son, presented what was modestly billed as a thirty-year vision: "This vision is designed with the time span of three hundred years. The next three decades is merely the first step." Another firm ostensibly embracing long-term thinking and expanding time is Rivian, an up-and-coming US-based electric car manufacturer. In a sense, they are trying to answer the question, How would one replace America's best-selling automobile—the Ford F-150 truck—with a more sustainable, future-proof version? Backed by, among others, Ford Motor Company itself and Amazon founder Jeff Bezos, Rivian's slogan is "forever"—suggesting its sci-fi-looking, climate-friendly SUVs and trucks are more in tune with the future of the planet. Whether a continuation of individual car ownership will be the best long-term solution for people, society, and the environment is another question.

Expansive thinking need not be the preserve of high-minded business leaders either. Governments are in an ideal position to think long-term since they act on behalf of an entire society. However, the compressed and ever-accelerating media cycle, combined with the electoral process that tends to focus more on short-term easy wins than on long-term hard choices, has tended to shorten the planning horizon for many public sector organizations. Popular pressure, driven by new digital media, has not helped either. Today we contend that our public organizations and our leaders face a serious challenge not only in terms of crafting long-term solutions, but also in terms of the very survival (long-term) of the democratic systems we have inherited.

Fortunately, some policy makers are starting to take the long view. For example, 2018 saw the launch of an All-Party Parliamentary Group on Future Generations in the UK. It seeks to "raise awareness of long-term issues; to explore ways to internalize longer-term considerations into today's decision-making; to create space for cross-party dialogue on combating short-termism in policy making." Likewise, Finland has had a Committee for the Future since 1993. A think tank for future science and technology policy, it seeks "to generate dialogue with the government on major future problems and opportunities."

Other countries have launched multiyear plans in response to climate change. For example, nineteen countries have signed on to the Carbon Neutrality Coalition Plan of Action—including, recently, Canada, Denmark, Spain, and the United Kingdom. Its members are committed to developing and publishing long-term strategies to achieve carbon neutrality in the second half of the century.

Meanwhile, the United Arab Emirates has embarked on a fifty-year plan to become "the best country in the world" in time for its centennial in 2071. And in 2012, Chinese President Xi Jinping referred for the first time to the "China Dream," or the elevation of China to become the world's economic, cultural, and military superpower. If the country follows the program Xi set out in his anthology of political theories, *The Governance of China*, it will achieve the "dream" by 2049—a century after the founding of the People's Republic of China. As Andrew Miller put it in *The Trumpet* in 2016, "Americans think in four-year election cycles. Chinese leaders think in terms of centuries."

Finally, there's the tiny Norwegian archipelago of Træna. Lying sixty-five kilometers from the mainland and home to fewer than five hundred people, it's one of the world's most remote communities. It's also the site of a remarkable cultural renaissance. A few years ago, the municipality decided to develop Træna into a sustainable community with versatile businesses, vibrant culture, and tourist attractions. At the heart of its drive to combine business, culture, tourism, and natural resources is its decision to take the long view. As Moa Bjørnson, Træna's head of development, puts it: "Humans have lived in this location for nine thousand years, so why not design a society for the next nine thousand?"

EXPANSIVE THINKING AND THE DESIGN OF EVERYDAY THINGS

We imagine that all this talk of long-term thinking might be making you feel giddy, so let's bring things back down to earth and consider one fairly elementary way in which this expansion could be applied.

In a previous book, *Instant Icon*, one of this book's authors, Jens Martin Skibsted, explored how companies could differentiate themselves from their competitors and in particular how the right product could provide the foundation for a company's business strategy. He pinpointed a product's lifetime as a key criterion. "Successful design products enjoy protracted lifetimes," Skibsted argued. "They remain relevant to customers for many years. A common way to express this is to call a product 'timeless.'"

So from a consumer perspective, one of the most attractive aspects of a product is that it potentially lasts a long time. But that's not all. Durable products are usually sustainable products.

"Fountain pens from Mont Blanc and watches from Rolex are the greenest products around," says the Australian inventor and entrepreneur Saul Griffith, founder of Otherlab. "Disposable digital watches which you only use for a short time and throw away with the battery inside are the worst. Products that people keep for a hundred years and pass on to the next generation can contribute to solving very serious environmental problems."

Griffith believes the design world is failing to find solutions to the environmental problems facing the world. For him, product longevity is one of the most logical places to apply input—and the intentionally long-lasting product provides an antidote to today's "use and discard" mentality.

Indeed, there's plenty of evidence of what's known as *planned obsolescence*: intentionally shortening the life span of a product so that its owner has to buy a new one. Most notably this is done via software updates that are in time incompatible with the hardware; by making the latest model irresistibly desirable to consumers; or by making its crucial parts especially fragile but expensive (if not impossible) to repair or replace.

According to the EU's Economic and Social Committee, the average life span of a household appliance two decades ago was ten to twelve years. Today it's six to eight years. We're also repairing old products less. According to Statistics Denmark, Danes spend 11 DKK ($1.75 USD) repairing electrical household appliances for every 1,000 DKK (approximately $160 USD) they spend on something new. And that's not because Danes find it easier to buy new. Thanks to planned obsolescence, it can be impossible to repair a product or replace a part. Little wonder that the average American owns twenty-nine mobile phones in his or her lifetime.

We believe firms should eschew the globally dominating model of designing for obsolescence and seeing value only in

EXPANSION 1—TIME

what they offer customers and shareholders. In particular, we believe the aesthetic appeal of Scandinavian design could provide a reference point for designers seeking to produce goods that will succeed in and benefit a more sustainable, circular economy. That is, if people are attracted to a product's look and feel—its durability and quality, heritage and craftsmanship—they're more likely to hold on to it, rather than lose, discard, or destroy it.

In fact, from Bang & Olufsen stereos to Arne Jacobsen chairs, Denmark has a long history of designing products that are meant to stand the test of time—quite literally. When Hurricane Hugo swept through the US Virgin Islands in 1989, many of the buildings erected over a century ago by the former colonial power, Denmark, were still standing, unlike later American-built ones. "The most beautiful houses on St. Croix, old buildings in the traditional style, have performed relatively well through the hurricanes because they are built for it," noted architect Ulla Lunn after visiting the island.

There's another backstop to short-term thinking in Denmark, and that's the dominance of family-owned or foundation-run companies. According to a 2010 study, family-run businesses in Denmark enjoy higher profit margins than non-family-run businesses in every sector but one (finance). More to the point, many of Denmark's family-run businesses have enjoyed global success. According to the EY and University of St. Gallen Global Family Business Index 2019, four of the world's largest family-owned businesses, ranked by revenues, are Danish: Grundfos, Lego, Danfoss, and AP Møller/Mærsk Group. (Only the latter is publicly held; the rest remain private companies.)

We believe the global success of these Danish businesses owes something to their establishment and ongoing existence as privately held, foundation- or family-run businesses. Unlike publicly floated companies, who are judged by the markets and always have one eye on the next quarterly report (to say nothing of shareholder activism) the secure structure of these large Danish companies means they can take a longer view of planning and do not have to be accountable to shareholders who may only be interested in making a quick return or challenge longer-term, more ambitious strategies. Take Mads Nipper, former CEO of Grundfos (and now CEO of Danish wind energy giant Ørsted). After he published a set of extremely ambitious long-term goals for the company's contributions to the UN Sustainable Development Goals, such as bringing clean drinking water to three hundred million people, he admitted to Danish Business Daily *Berlingske*: "If this was a publicly listed company, shareholders would have me fired."

As firms are increasingly expected by their wider stakeholder landscape to make real contributions to global challenges, inspiration can be drawn from the long-term perspective and more holistic thinking of foundation-owned firms. Our point is this: even publicly listed companies can choose to take the long view, if they wish to; but currently, it may well be at their peril due to the short-term thinking of most shareholders. The shift to long-term thinking can thus come either from shareholders who start demanding a more strategic, longer-term perspective from the firm's boards and C-suites; from bold leadership within the companies themselves; or both. We believe a sustainable future for business requires expansive thinking, and the more pressure the surrounding share- and stakeholder landscape puts on firms to take the long view, the better. If we can create a new framework for businesses to pursue long-term goals, we might also see businesses solve long-term challenges.

Some stories are too good to be true—including, alas, the legend of the New College oaks. While the beams in its dining hall were indeed replaced in the mid-nineteenth century, using oaks growing in forests that New College owned, those trees hadn't been planted for the express purpose of replacing the beams. For one thing, New College had only owned the forests since 1441—sixty years after the hall was built—and it had long been standard woodland management to grow oaks alongside hazel and ash trees in the forest.

Even so, the New College oaks were occasionally used for major construction projects, such as shipbuilding. As the website Atlas Obscura has concluded, "While the story is perhaps apocryphal, the idea of replacing and managing resources for the future, and *the lesson in long-term thinking is not*" (emphasis ours).

Indeed, whether the New College oaks were intended to be used to replace the beams is beside the point. They were planted in the expectation that eventually, hundreds of years in the future, they would be used for major construction work. And none of the people who planted the trees believed they would live long enough to enjoy their benefit.

Taking the long view is thus about collapsing the distance between now and then, today and tomorrow, this century and the next. In that respect, taking the long view is a matter of proximity—the subject of our second expansion.

3

EXPANSION 2– PROXIMITY

The traditional view of the sanctity of human life will collapse under pressure from scientific, technological and demographic developments.
Peter Singer

A value is like a fax machine: it's not much use if you're the only one who has one.
Kwame Anthony Appiah, *The Ethics of Identity*

Even to Danes used to dreadful weather, the sight of twelve blocks of ice in Copenhagen's Town Hall Square must have come as a shock. Fortunately, it wasn't the result of freak weather, but a daring new work by the Danish-Icelandic artist Olafur Eliasson in collaboration with geologist Minik Rosing. They had harvested free-floating chunks of glacier ice that had broken off the ice sheet due to global warming and shipped them from Greenland to Copenhagen. Then, they had arranged them in a circle in the heart of the Danish capital, where they slowly started to melt.

Dubbed "Ice Watch," the 2014 installation was designed to make Copenhageners think about climate change. The symbolism of the melting icebergs—arranged to resemble the dial of a watch—is impossible to ignore. Time is running out, it says. The clock is ticking.

Yet Eliasson's eye-catching installation does two other interesting things. One is that it collapses the distance between people and the Arctic. "Out of sight, out of mind" runs the adage. "Not so fast," responds Eliasson, as he deposits tons of glacier ice in city centers across the globe. Second, in its pops and cracks—audible signals from the frontline of climate change—the installation collapses the enormous distance between our planet's past and present and, thus, its present and future. In this way, "Ice Watch" intentionally humbles us—and sounds the alarm bell.

Describing how he conceived of "Ice Watch," Eliasson told us he had been following the work of the Intergovernmental Panel on Climate Change (IPCC), which monitors the well-being of the planet and greenhouse gas emissions. He says he realized that they were producing a "data-driven narrative"—which sparked his interest in "how to make data tangible, how to give a physical, emotional narrative to the data." Knowing that "if you combine sensory, firsthand experiences with knowledge-based, cognitive experiences, you are more likely to find behavior change," he set about creating a tangible work of art.

Eliasson believes the climate crisis requires expansive thinking. "We mismanaged the planet on which we are sitting. We made the mistake of taking for granted that resources are unlimited, that the climate crisis would solve itself, or that some genius engineer would come up with a push-button solution," he told us. "We also failed to come up with an economic system that will have built in a reaction model that would take the challenge of paying for the climate crisis. So this means a lot of the models we rely on today which are derived from success criteria of the past are very hard to adapt to the future."

Eliasson's urgent artwork embodies the expansion discussed in this chapter, which concerns distance. Geographic distance. Topological distance. Distance between people. Distance between people and our planet. Distance in terms of time. In other words, this expansion concerns what we call proximity—and invites the reader to rethink what's perceived to be "close" or "far away."

We begin the chapter with a short philosophical discussion—and an invitation to reconsider what we mean by "proximity." Our aim is to demonstrate that thinking differently about proximity could lead to more ethical behavior, by which we mean having a greater feeling of care or empathy for other people—if not other species and the planet per se. This shift in thinking, this expansion, underpins the entire chapter.

EXPANDING "PROXIMITY" THROUGH ETHICS

Picture the scene. You're reading the newspaper when you spot an ad for an antipoverty charity, announcing its latest fundraising campaign. It uses a photograph of a poor African family. Though it tugs your heartstrings, you turn the page. Later, on the way to work, a beggar asks you for money. Instinctively, you give him some spare change. Why?

Before you finish answering the question, you spot a group of commuters laughing and jostling on the platform. As you get closer, you realize they're fighting. In fact, one of them is badly outnumbered. You decide to break up the scuffle before someone gets badly hurt. Later, you reflect on the fact that the victim looked like you—same age, sex, ethnicity, and class. Is that why you decided to get involved and come to his aid, you wonder?

When you return home that night, you realize you forgot to take out the trash. A fly is buzzing around the room. Meanwhile, your cat is pestering you for its supper. You're irritated by both the cat and the fly. Although it never crosses your mind to harm your cat, you have no qualms at all about rolling up your newspaper and swatting the fly. Why?

A simplistic thought experiment, these scenarios reveal the potential parameters of our sense of "proximity" to other people—and, indeed, other species. For various reasons related to age, gender, and cultural and social backgrounds, we all have different ethical beliefs—which directly inform our feeling of closeness to or kinship with other people.

These beliefs explain why we decide to help the local beggar but not starving families overseas, why we intervene to stop squabbles on our doorstep but turn a blind eye to conflict farther afield, and why we ensure the rights and welfare of pets but not bugs.

To understand what we mean by "proximity" or "perceived distance," and how to expand it, we first need to accept that it has an ethical dimension—in other words, that we need to rethink our ethical outlook. We believe that "Actics," a paper written by Danish philosopher Janus Friis in 2004 and based on the work of Nicolai Peitersen and Jens Martin Skibsted, provides a useful tool for doing so.

A combination of action and ethics, "actics" is an ethical system concerned with the actions of agents—meaning someone or something able to act ethically, including people, companies, and nations. Ethical behavior is about adding value to the world and is sparked by an ethical feeling of care, and care is a feeling of proximity to someone or something. The logic is thus: expand your feeling of proximity to others and you trigger greater ethical behavior and, in turn, add more value to the world.

Proximity can be broken down into a number of categories, including social, ethnic, genetic, cultural, physical, religious, ideological, and sexual. Each category can then be viewed on a sliding scale, from "close" or "alike" to "distant" or "unlike," with our chosen position on the scale a reflection of our perception of proximity to another person (or place, species, etc.).

The problem, so to speak, concerns where we place ourselves on those sliding scales: Do I feel close to someone with a particular characteristic, or don't I? Remember the argument presented by *actics*: ethical behavior is about adding value to the world. Such behavior mainly derives from having a feeling of care, which in turn comes from feeling close to someone or something. Our aim, then, is to care more: to expand our feeling of proximity, to shift our position on those scales, to increase the amount of value we create.

Fine in principle, you might say, but how to go about it in practice? We can start to care more either by gradually widening a certain category of proximity (so we relate to it more easily) or by taking a determined leap in proximity (putting ourselves into a radically different context and thereby empathizing with those living in it). Proximity is established by ensuring that the solutions created address actual problems close to you that you care for.

This chapter explores ways in which our notion of proximity is changing. It begins by considering the benefits—even the necessity—of reconsidering what we see as "close" to us in a geographical sense, in the light of climate change and the accelerating need for innovation. We then shift to how technology is making the ground shift beneath our feet, collapsing notions of time and distance that we have always taken for granted.

EXPANDING GEOGRAPHY

What is proximate is currently expanding in the simplest way: what is close geographically from the point of urgency but also of innovation is getting closer. Let's consider a continent that was not only the cradle of mankind but that we would be wise to develop a closer sense of proximity to.

The largest future population growth will happen in Africa, necessitating design tailored for minimum energy use and sustainability, including new products and services. Climate change means famine, drought, flooding, and increasingly volatile weather will become routine occurrences. The best solutions to these problems will dominate the conversion to a green economy, as indeed they did during the 2021 COP26 climate summit in Glasgow. But unlike Western democracies, African countries can't afford to engage passively with the challenges. The continent will face the brunt of climate change.

Indeed, the countries affected first and most severely will have to move fast and adapt. Radical innovation typically gains traction at the margins of the market, which is why we should be looking to Africa if we want to stay ahead of the curve. Indeed, the challenges Africa faces are already unleashing an innovative and entrepreneurial spirit that could well set the pace for the global design scene. Design is booming on the continent, with both necessity and a relative lack of common red tape driving innovation.

Take the vernacular-architecture movement, spearheaded by a young generation of stars such as Issa Diabaté, Francis Kéré, and Christian Benimana. It is working on environmentally friendly, minimal-resource buildings based on modernizing traditional construction techniques. Meanwhile, if you're looking for smart water innovation, head to South Africa, for instance, where Drybath Gel can replace a shower and save gallons of water—an absolute necessity during Cape Town's extreme water shortage of 2017–2018. For off-grid innovation, you could look to Kenya and the M-Kopa solar battery/Wi-Fi/LED lamp/business hub. Or at how Zipline has pioneered drone delivery in Rwanda and Ghana.

Make no mistake: we are fast moving toward a brave new world of design-led climate adaptation, one that requires us to place ourselves as observers, or even as entrepreneurs, in proximity to the bleeding edge of the action. And as this action is shaped by how we adopt technology, the concept of geographical proximity is converging with topology.

EXPANDING WHAT WE MEAN BY "CLOSE" AND "FARAWAY"

"If you want to send a message, use Western Union," the movie mogul Samuel Goldwyn supposedly said about films he deemed excessively preachy. His quip would be lost on many people now. After all, the telegram has gone the way of the rotary telephone. WhatsApp would be the right reference today. Or perhaps Skype, Signal, Zoom, or any modern communications tool that has collapsed the distance between us. Holographic technology could be the next game changer, impacting everything from interpersonal communications to first aid, real estate to manufacturing.

EXPAND

In *The Seventh Sense: Power, Fortune, and Survival in the Age of Networks*, Joshua Cooper Ramo discusses the difference between two ways of measuring proximity—geography and topology—and assesses the consequences of the latter changing in a flash.

Geography is concerned with the fixed distance between two points. Topology expresses the idea that the speed and distance between two points affects how "distant" they are. For instance, Los Angeles and New York will always be about 2,800 miles apart. In topological terms, however, they are as "far apart" as the fastest connection between them—which is to say, right now, about 0.3 milliseconds along a high-speed fiber-optic cable.

"Geographies are pretty much constant; topologies can change in an instant," Ramo writes. And changing topologies explain the existence of much of what we understand about modern life, including the World Wide Web, computer hackers, financial market "flash crashes," underground terror cells, and commonplace communications tools used by modern-day movie moguls to send messages.

The point isn't just that topologies are changing or that they are changing as fast as technology is. It is that changing topologies in and of themselves change other topologies and make the world even closer. "Every new piece of a network, every new platform or protocol, alters how we connect," Ramo writes. "This process works on our sense of distance like an efficient, strange sewing machine: Something very far away can be, suddenly, with one stitch of innovation, right on top of you. The speed and the quality of a connection are what determine how honestly 'near' and 'far' something is."

And what this gives rise to is a concept Ramo calls *locational utility*, meaning that even if something or someone remains the same physical distance from us, the "closer" it gets in terms of increased connection and speed, the more useful or valuable it becomes to us. In other words, technology has upended and expanded our understanding of "distance." And to see what this means for design, you need only look at its impact on the world of manufacturing.

EXPANDING "THERE AND THEN" TO "RIGHT HERE, RIGHT NOW"

Traditionally, if you needed a new desk for your office, you had three options. You could buy a brand-new desk from a shop or a used desk from a stranger. You could borrow one from a friend or inherit one from a relative. Or you could pay a cabinetmaker to build one for you. But no matter which option you chose, you would have to go and get the desk or have it delivered to you. And if you liked the look of a particular desk in, say, Milan or Tokyo, you certainly had your work cut out.

Digital fabrication threatens to disrupt all that. In theory, it allows anyone, anywhere, to physically make or reproduce objects that would ordinarily require specialist manual skills, if not industrial-grade machinery. The stock-in-trade tools of digital fabrication are 3-D printers, laser cutters, and CNC machines, which are capable of "reading" digital files and cutting, carving, and shaping objects with pinpoint precision out of practically every manufacturing material under the sun, from titanium to cement.

Moreover, these machines are getting smaller, cheaper, and more accessible, leading to predictions that "soon our factories could sit in our homes or on street corners," making local production the new normal and ensuring that "everything from clothing to furniture could be downloaded, customized and produced as needed." This shift in the proximity of manufacturing to where we live, work, and play is already shaping new business models and ways of creating value.

Consider two examples of the digital fabrication revolution. Paperhouses is an open-source platform that lets people freely download PDFs of blueprints drawn up by leading architects—allowing them to digitally manufacture site-specific buildings. And Opendesk is a London-based company that enables people to download its designs and make items of furniture locally using a CNC machine. In other words, the company is aiming to cut out the middlemen and logistics of the traditional global supply chain and pioneer a new kind of supply chain and business model. (It even designs office desks.)

The idea of building a new model of on-site local and sustainable manufacturing is attractive to cities too as they endeavor to bring back high-value-added jobs in an urban context. A fast-growing network of cities that call themselves "Fab City," led by Barcelona, are striving to become such global hubs of localized manufacturing.

In the late twentieth century, we thought of manufacturing as something that was far away—noisy, polluting factories that needed to be built well beyond our cities. We also built complex and far-reaching value chains that, say, allowed something to be designed in Britain, manufactured in China, and sold in the United States. The advent of new digital manufacturing technologies enables a shift in the proximity of industry. A shift that invites us to expand our perception of how and where we can produce things, and how manufacturing—green, sustainable, and tailor-made—can yet again become part of our urban life.

EXPANSION 2—PROXIMITY

EXPANDING CARE FOR OTHER (HUMAN) BEINGS

Rapid developments in the digital realm have expanded our thinking about "distance" and, with it, our notion of what's "close." Digital fabrication tools and open-source manufacturing expand our opportunities to design, produce, and use materials. The third proximity-related expansion concerns people—and builds on the ideas about ethical behavior, which we explored at the beginning of the chapter.

The world is coming apart at the seams, and we need to find not only ways to mend it, but also the motivation to get started on a new path and the emotional guidance to steer us along the way. Could empathy be the answer? This section concerns the ethical potential of collapsing the distance between us, literally and figuratively. For instance, can we design ways to reduce the empathy gap so that we care more for, say, migrants trying to make it out of the Global South and into the rich economies of Europe and the United States?

Adam Smith wrote in *The Theory of Moral Sentiments* that most people would be more perturbed by the loss of a half a finger than if one million Chinese people perished. In theory, high-speed fiber-optic cables puts China a few keystrokes and seconds away from us. Yet for a long time, the news that a million Chinese Muslims are in internment camps—called vocational education and training centers—scarcely registered. So how do we close that gap?

Could altering reality itself, through virtual reality, provide an answer? A new experience at the Red Cross's headquarters in central Copenhagen aims to put people in other people's shoes—literally. The first thing that visitors do is put on a pair of rubber boots, as if preparing to enter a refugee camp. The inner soles are filled with a gel that mimics the sensation of trudging through mud, a further attempt to collapse the distance between the visitor and the hardship.

Visitors then proceed through a series of six immersive environments, each designed to showcase the kind of work that the Red Cross does. Authenticity is the watchword. From the tarpaulin tent of a field hospital in South Sudan to the asphalt road surface used to demonstrate the organization's work training people in emergency first aid, all materials are real. Visitors also hear real people's voices—refugees, victims of domestic violence—and wear VR headsets that make them witnesses to Red Cross food drops or caravans of refugees.

"It's more motivating and more engaging when we use authenticity and as many senses as possible," says Morten Schwarz Lausten, senior communications advisor at the Red Cross. As many as twenty different voices are used in different ways to make the experience more humanizing: "The voices of real people are so important."

Some five thousand people are expected to visit the new Red Cross Experience annually, many of them teenage pupils from Danish schools, but also Red Cross volunteers and corporate backers. The experience center is an attempt to bridge the gap and make the abstract more real—to expand people's understanding of why the Red Cross does what it does. "There's this sense of the big why," Lausten says. "Why do we help people? We are always looking for strategic communication methods and tools to engage. And all our communication should create empathy. That's the primary purpose. We have to give people in need a voice, to create empathy."

As part of the Red Cross experience, visitors are invited to play several "games" that involve moral dilemmas typically faced by refugees or Red Cross employees in the field. For example, one game puts the participant in the position of a Sudanese refugee who has saved enough money to buy a spot in a smuggler's boat, only to discover that it is overcrowded and in poor shape. What should he do? Another game requires visitors to match wooden blocks representing displaced people, a task made trickier by the unnerving audio of strafing Syrian fighter jets.

One thing is closeness to refugees; another is closeness to war. "The Enemy," a French VR installation, immerses users in three conflict zones, allowing them to "interact" with real-life combatants in Salvador, the Democratic Republic of the Congo, and Israel/Palestine. The combatant's testimonies and confessions about their lives, experiences, and perspectives on war allow people to better understand their motivations—and accept their humanity.

In the same vein, Italian data-art firm Accurat, which creates stunning visualizations and virtual and augmented-reality applications, has collaborated with Google for the project "Building Hopes." This interactive visual tool invites users to select global themes for which they are hopeful—such as learning from history, human rights, or humans on Mars, and connect with an augmented-reality visualization of online data that relates to the given theme. Accurat describes this as "an immersive data-art experience that invites people to materialize their hopes as permanent augmented reality sculptures and to use them as a lens to explore and understand Google Trends data."

Gamification is a tool that many organizations are exploring online. In recent years, both the BBC and the *Guardian* have developed interactive games designed to show people the experience of Syrian refugees trying to make their way to Europe. And the development of television programs such as *Bandersnatch*, a feature-length episode of the successful Netflix series *Black Mirror*, featuring a choose-your-own-adventure narrative that requires viewers to make on-screen decisions shows how this emerging style of storytelling could be employed to motivate, engage, and introduce people to complex topics—and potentially bridge the empathy gap.

Digital proximity might trigger radically more ethical behavior. Its potential is largely untapped by Silicon Valley–type digital ventures and Big Tech, and it could add massive value on a global scale.

All of these approaches serve to expand the participant's feeling of proximity—their feeling of closeness to problems and to those experiencing them. This matters to ordinary citizens in shaping their understanding of pressing societal issues, but perhaps even more importantly, it can engage decision makers to respond with more empathy.

EXPANDING—IN ORDER TO COLLAPSE THE DISTANCE BETWEEN PEOPLE AND POWER

Expanding our concept of proximity doesn't just mean generating more closeness and empathy between people. Another consequence could be that it helps reduce the gap that has long existed between people with power—politicians and elites, broadly speaking—and ordinary citizens.

"Elites, who define the issues, have lost touch with the people," wrote Christopher Lasch in *The Revolt of the Elites and the Betrayal of Democracy*. "The new elites are at home only in transit, en route to a high-level conference, to the grand opening of a new franchise, to an international film festival, or to an undiscovered resort. Theirs is essentially a tourist's view of the world—not a perspective likely to encourage a passionate devotion to democracy." That was more than twenty years ago. Does anyone seriously doubt that today's political upheavals are anything but a partial consequence of the political classes' estrangement from the people they purport to lead and represent? From MAGA-cap wearers to the rejection of European political harmonization and multilateralism, to the Yellow Vests antagonized by Macron's suggestion of "crossing the street" to get a job, populism is on the rise and revolting against elitist rule.

Where might we be today if power weren't so clustered and cloistered, if elites weren't so "far" away from regular voters? One could also argue that this loss of identity necessarily depletes feelings of proximity toward others and diminishes the subsequent compunction to care for them—further entrenching political positions and creating emboldened in-groups and out-groups. Most of us come from somewhere in particular and engage with other people from that perspective—but the smaller that "somewhere" feels to us, the less proximate "elsewhere" seems—and the less we care for it or the people who come from it. Certainly, if we want to save the planet, we need to do everything we can to reverse this tide and expand our sense of proximity toward others.

All of which begs the question of how might we reduce the gap or shrink the distance between elites and ordinary people. One way might be if more politicians and decision makers were more connected to the lived experience of voters. Another might be to try to make more connections between people, to foster their capabilities, and to create beneficial relationships.

A decade ago, the British social entrepreneur Hilary Cottam set out to understand what young people needed from public services. She showed up at night on what she called "rough ground"—London estates and seaside towns—with a rented bus stacked with pizza.

"I wanted to listen and to think about what could be," she said. "What I heard again and again was a hunger for connections, for the relationships needed to get on."

Likewise, in her recent book *Radical Help*, Cottam describes an experiment undertaken in Swindon, a town of London whose council was struggling to cope with "chaotic families," some of whom had as many as seventy-three different professionals involved in their lives. Cottam says she simply sat down with some of the families and asked them what changes they would like to see to help to turn their lives around. Her approach, which had tangible results, had a deep human connection at its heart—and one that was only possible because of a shift in proximity. What if more political decisions were based on this degree of closeness, insight, and empathy?

EXPANDING THE FUTURE OF DEMOCRACY BY INCLUDING CITIZENS IN POLICY MAKING

Does a revitalization of the proximity between elites and people and communities in crisis require mainly such local, bottom-up approaches? Or might a redesigned interaction between political processes and citizens also start at the top, with redesigned political and policy-making processes? After all, the utopian democratic system is one of accountability. Discontent with the performance of elected officials leads to their replacement through elections in a competitive environment, rewarding those legislators who provide the best solutions to society. Ideally. But as we've seen on numerous occasions in recent years, politicians are being elected on populist agendas that fail to take seriously many of society's most complex challenges—including the climate crisis, globalization, declining living standards, and job-market insecurity.

Each time the narrow sightedness of elected leaders pushes a nation in the wrong direction, the challenges for the rest of the world grow. Democracies have to rise to the occasion, and design will and should play a role. The modern democratic states face two challenges: one is to build better government systems that increase trust and understanding between the state and the citizen; the other is to start rethinking the inner workings of modern democracies to ensure that they deliver real results for people in their immediate everyday lives and their future.

In the past decade, a number of governments have approached problem solving by creating, at local, regional, and national levels, innovation and design labs that apply a design methodology to policy making and implementation. For instance, in 2001, the Danish government set up MindLab, an internal unit reporting to a handful of ministries, with the mandate to involve citizens and businesses much more directly in the development of new and innovative public policies and services. (Christian led MindLab from 2007 to 2014.) Under this approach, policy can be informed based on synthesizing disparate perspectives with a view to creating more relevant, citizen-centered policy responses through iterations and by ensuring collaboration between the affected and the legislators. Over the past two decades, such labs have proliferated across the planet and can be found in both advanced and emerging economies.

In public innovation labs, the end product has to be compatible with public, not private, bottom lines. Public value could be things such as productivity, service, changes in behavior, and democracy. How does each policy contribute to society within these criteria? Legislation is a product of policy making, so why shouldn't it be prototyped and tested at local levels to ensure that it is actually addressing the right problem?

While the Danish MindLab has since been discontinued, the labs movement is very much alive as a response to the need for more and more relevant proximity between decision makers and constituents. The government innovation-lab approach has gone global, but we're still at a point where we need to further develop the combination of design and policy. Raymond Loewy once said that design is too important to be left to designers, but the same goes for policy making: it shouldn't only be left to civil servants, and designers can help in finding the right problems and in developing better solutions—not least by bringing citizens into the process at the earliest stage.

The question remains, however: Does democracy suit the needs of the twenty-first century? Today's democratic systems, as well as their mechanisms, were defined centuries ago. The American constitution was crafted in the late 1700s, when there were no automobiles, nuclear weapons, precision medicine, or social media. Denmark's constitution, which dates to 1849, is not much more modern. According to the Constitute Project, an open-source repository of national constitutions, few countries crafted their constitution in the latter part of the twentieth century or at the start of the current one. Even those who wrote new constitutions—usually out of necessity, such as many post-Soviet countries—did so using the blueprints of age-old Western ones.

Within the next decade, we'll probably only see more suggestions for solutions to democracy's problems. We're already seeing ideas surface, not least in regard to voting. For example, the idea of liquid democracy proposed by the nonprofit *Democracy. Earth* allows the secure transfer of votes to others based on that person's expertise or merit. We could also see proposals for dealing with how often we vote, in which regions our votes should count, better ways to carry out electronic voting—and perhaps even more radically, whether certain topics should be moved from the political to the technocratic sphere. Just as monetary policy is the domain of the Federal Reserve or the European Central Bank, could we imagine a future in which questions regarding the climate crisis were treated supranationally by a technocratic climate board?

Visionary architect Bjarke Ingels thinks so. "With the right group of companies, scientists, holistic designers, and thinkers, we could create a master plan for the planet—call it the master planet—that would be much more tangible and actionable than this kind of endless stream of [political] declarations," he told us. Just as one wouldn't want to build a cathedral without blueprints, nor should we try to stop climate change without a similar kind of master plan. "We are at a juncture where we have the necessary tools and the necessary knowledge. We can, with the right team, create the master plan for the planet that would vastly increase our chances of success solving the greatest problem we have ever had as a species." Is this the moment for Plato's philosopher kings to become the ultimate decision makers?

Responding to the gravity and urgency of the climate crisis challenges the limits of modern democracies. The near future will determine how democracies respond and adapt. Will they act in elitist ways and thus prove themselves at best dysfunctional or at worst obsolete? With their current design and trajectory, democracies are likely to be increasingly challenged. That's why the work of redesigning a modern democratic, responsive, and agile state should start today. We must, in this sense, politicize design and start designing for more inclusive political decision-making.

EXPAND

EXPANDING PROXIMITY TO DESIGN FOR INCLUSION

We began the chapter with a short discussion of actics, a philosophical theory that makes a number of compelling arguments, among them that proximity can be broken down into a number of categories—social, ethnic, genetic, cultural, and the like. We suggested that we place ourselves on a sliding scale for each category according to our sense of proximity to a person, place, species, and so on, and that our aim, as ethical human beings, should be to care more, empathize more, expand our sense of proximity, and thereby transform how we create value in the world. What is valuable is not objective. What we as humans consider valuable is always (also) an ethical question. To embrace proximity as an expansion itself is an ethical marker.

Actics argues that we can start to care more either by gradually widening a certain category of proximity or by taking a determined leap in proximity, expanding the stakeholder group, so to speak. But what if the world were designed in a way that stopped us from doing so? What if there were roadblocks preventing us from taking the kind of leap we need to make? What if those roadblocks were inadvertently—unconsciously—put in our way? And what if knowing they were there meant we could do something about them? Could current ethical approaches and tenets be as antiquated as our democratic practices?

That, essentially, describes the problem with much of the way we've historically designed the world—and, in the case of emerging technologies such as artificial intelligence, are still designing it. Expanding "proximity" concerns who designs the world and who they design it for. In this sense, it's a question of reducing the gap between the designer and the product, the designer and the end user—whoever or whatever that may be.

Let's consider some examples of the roadblocks that prohibit us from expanding our proximity and how those barriers to ethical behavior can have adverse consequences. While challenges of proximity and ethical design are present everywhere, the rise of big data and use of AI and machine learning has heralded a new era of potentially biased design with more speed and scale than ever before.

First, consider gender. In her recent book *Invisible Women: Exposing Data Bias in a World Designed for Men*, the British feminist writer and campaigner Caroline Criado Perez launches a broadside against the world of design. "Designers may believe they are making products for everyone," she argues, "but in reality they are mainly making them for men. It's time to start designing women in."

To make her case, Criado Perez marshals a wealth of evidence for the existence of a "gender data gap"—an absence of information about the impact of a product or service on women. "Films, news, literature, science, city planning, economics . . . are all marked—disfigured—by a female-shaped 'absent presence,'" she writes. "The gender data gap is both a cause and a consequence of the type of unthinking that conceives of humanity as almost exclusively male."

In some cases, a world designed largely by men for men is irritating but harmless. The top shelf is set at a male height norm. Men's and women's lavatories are given identical floor space, even though women take longer to use them. The typical A1 architect's portfolio fits under most men's arms, while most women can't reach theirs around it. And to widespread derision, a company designs a ballpoint pen "just for women," replete with a "sleek silhouette, jewelled accents," and "a soft contoured grip for all day comfort."

In other cases, however, the gender data gap has graver consequences. In March 2019, NASA was set to make history with the first-ever all-female spacewalk. The milestone moment had to be postponed, however, because there was only one safe-to-use and ready-to-wear spacesuit in the correct size for both women on the International Space Station. It doesn't bode well for our plans to colonize Mars if we can't properly dress half the people who would be making the trip.

Meanwhile, in New York State, women who work as carpenters have reported higher rates of wrist sprains and strains, most likely because "standard" construction-site equipment is designed for men. Worse, crash-test dummies have long been based on the "average" male body, and car safety tests haven't accounted for women's measurements. "When a woman is involved in a car crash, she is 47 percent more likely to be seriously injured, and 71 percent more likely to be moderately injured, even when researchers control for factors such as height, weight, seatbelt usage, and crash intensity," Criado Perez writes. "She is also 17 percent more likely to die. And it's all to do with how the car is designed—and for whom."

In *Weapons of Math Destruction*, Cathy O'Neil offers another grim example of inherent gender bias. "An engineering firm wants to hire an engineer, but in order to build an algorithm to help it, it needs to define success," she writes. "It defines success with historical data as someone who has been there for two years and has been promoted at least once. The historical data says no woman has ever been here for two years and been promoted, so then the algorithm learns that women will never succeed."

EXPANSION 2—PROXIMITY

Second, consider how questions of ethnicity are often ignored in technology design. Of course, one day we'll see self-driving cars on the streets, so this thorny problem won't matter anymore, right? Maybe not, but the dawn of the driverless vehicle only brings another horror show into focus. A study from the Georgia Institute of Technology found that a pedestrian with dark skin is more likely than a person with a lighter skin tone to get hit by a self-driving car—all because automated vehicles may be better at detecting the latter.

The study is by no means an outlier. Sadly, there is a growing body of evidence of human bias creeping into automated decision-making systems like the ones in driverless cars. It's called algorithmic bias, and it affects everything from Google image searches to police profiling systems to credit ratings and job applications. As many industry observers point out, these tools are never neutral but in fact reflect the ideology of those who make them.

The applicable acronym is GIGO: garbage in, garbage out. In other words, the algorithm is only ever as biased as its creator. Which brings us back to proximity. The bigger the gap between the creator of the algorithm and the people that algorithm impacts, the bigger the potential impact that that algorithm has. If a driverless car is "trained" to see lighter-skinned pedestrians as objects to avoid, it will avoid lighter-skinned people more successfully than darker-skinned people.

Algorithms also determine what advertising we see online and what stories appear in our Facebook newsfeeds. That shapes our supposedly "original" views and tastes, and builds and reinforces cultural and ideological "filter bubbles" around us. Technological shifts and changing topology may have collapsed the distance between us, but they've also made it easier for people to be less proximate to one another. In the age of the algorithm, proximity may be an illusion.

The solutions are obvious but bear repeating. First, the data could be better. In the example earlier, designers could use more darker-skinned people to train self-driving cars, and weight the sample of darker-skinned people more heavily too. Second, the team of designers itself could be more diverse—or in keeping with this expansion, they could be more proximate to the problem they're trying to solve. Put bluntly, a minority designer is more likely to check whether the driverless car he or she is designing is falling foul of any biases that could negatively impact minority pedestrians (or indeed drivers). Basically, you need to design for the people not in the room or invite them in.

Third, let's consider expanding our notion of proximity beyond human characteristics such as gender and ethnicity. We need not limit our expansion in thinking about proximity to our own species. Rather, in its most ambitious sense, doing so would see us rethink or refocus design so that it isn't just human-centric or humanity-centric, but life-centric or planet-centric. Having seen a shift toward human-centered design, might it be time for a further expansion into life-centered design?

That's certainly the thinking behind the burgeoning "rights of nature" movement, which promotes the adoption into national legal systems of laws recognizing such rights. According to the Global Alliance for the Rights of Nature: "Under the current system of law in almost every country, nature is considered to be property, a treatment which confers upon the property owner the right to destroy ecosystems and nature on that property. When we talk about the 'rights of nature,' it means recognizing that ecosystems and natural communities are not merely property that can be owned, but are entities that have an independent *right to exist and flourish.*"

Both Bolivia and Ecuador have implemented laws that recognize certain rights for nature. For instance, according to the Global Alliance for the Rights of Nature, while Ecuador's 2008 constitutional change did not automatically protect nature, it did give citizens the legal authority to enforce these rights on behalf of ecosystems. In other words, the ecosystem itself can be named as the plaintiff.

In terms of how we might design the world differently, this is nothing short of revolutionary. After all, that onetime darling of the C-suite, design thinking, assumes there's always a user, but what about this plaintiff, the ecosystem? How is an entire ecosystem a user? When the actual "user" is the culprit, the defendant on trial, we must think beyond user-centric design—and contemplate "life-centered" design instead.

Some interesting projects in this field are underway. The Copenhagen Institute of Interaction Design (CIID), for example, is investigating life-centered design in Costa Rica, whose rich biodiversity hosts more than 5 percent of the world's species. The institute believes that "designing for inclusion and empathy is an absolute necessity in today's world" and has started offering intensive workshops in Costa Rica where participants will "dive deep into collaboration and explore how working with our differences can create meaningful experiences," using the framework of the UN Sustainable Development Goals.

"At this point in our global ecological crisis, the survival of humanity will require a fundamental shift in our attitude toward nature: from finding out how we can dominate and manipulate nature to how we can learn from it," the CIID explains. "This

EXPAND

course could be summed up as an awakening: reconnection with Nature of which we are part."

A great example of a project developed by CIID students is the BioCollar, a wearable electronic device that "acts as an intermediary between plants and humans" and aims to build empathy—and thus increase the sense of proximity—between the plant and its owner. A device in the plant pot connects wirelessly with the collar, which the owner wears around their neck, allowing them to get real-time information about their plant's needs. For example, if the plant requires water, the BioCollar tightens (slightly!). If the plant is getting too much sunlight, the collar gets warmer. And if pests infest the plant, the collar vibrates.

Of course, BioCollar is a primitive prototype. But it isn't hard to imagine its potential—and the potential of devices like it—in the years to come. CIID's trailblazing course—and others like it—would be noteworthy under any circumstances. It takes on greater significance, however, in light of our final proposal for expanding how we think about proximity.

Designing for inclusion means incorporating more dimensions of proximity into the process of creating new solutions. We've illustrated this through examples of how our current biases skew the design process leading to adverse outcomes in digital solutions, whether in terms of gender or ethnicity, or in the question of designing for people or the planet. The first step, in all these instances and others, is awareness: to ask new questions of proximity and use "tools of empathy," including those offered by immersive experiences and virtual and augmented reality, to create empathy, reducing the distance between people and problems, between those who design and those who are designed for.

EXPANSION 2—PROXIMITY

EXPANDING PROXIMITY TO SAVE THE PLANET

A year after he hauled twelve chunks of glacier ice from Greenland to Copenhagen and arranged them in a circle in the city center, Olafur Eliasson repeated the display in London and Paris. The *New Yorker* magazine was on hand—and Eliasson gave them an indelible image. If passersby put their ears to one of the icebergs, they might hear a pop, as air trapped in the ice fifteen millennia ago escaped into the atmosphere. "It is a little pop that has traveled fifteen thousand years to meet you in Paris, and tell the story of climate change," he told the magazine.

In this chapter, we have suggested several ways of rethinking proximity, including what we mean by "distance," the consequences of this for how—and for whom—we produce and design goods and services. We have argued that we ought to care more for other people—and that reassessing our notion of proximity toward others is where we need to start.

In a sense, this expansion concerns urgency because it is explicitly calling for a greater sense of empathy and proximity with people and places, and even with nature and our planet as a whole. For every decision we make about what to create, we should consider whether it is ethical and inclusive in the most expansive of senses.

In that respect, it is a callback to our first expansion, which explored time and called for more "long-term thinking." In that chapter, we met some inspirational thinkers who want us to think differently about time—like the brainiacs at the Long Now Foundation, who are building a clock that will run for ten thousand years and help foster long-term thinking.

We also argued that, instead of designing for the future, we tend to design for the here and now. And we pointed the finger at speed, warning that speed of product development all too often means that traditional design phases such as ideation and research are truncated, that iterations become the implementation, and what used to be beta phases are accepted as end products.

In that sense, the expansion of time and the expansion of proximity are opposite sides of the same coin. Just as in chapter 1 we suggested expanding our thinking so that we can conceive of a distant future, this chapter has asked us to expand it so that we can conceive of seemingly distant people and places and events. This is really two ways of thinking about distance, one being distance in time, the other distance in empathy, and both are about thinking differently in order to compress those gaps.

So what is the urgency? It is that we don't have the luxury of time. Faced with the potential of cataclysmic climate change, speed really is of the essence. Humanity faces an existential threat. But we aren't doing enough to prevent it. The warnings are dire. Until the COVID-19 pandemic shuttered schools for months and made protests unsafe, a generation of schoolchildren started walking out of the classroom and going on "climate strike" as part of the Greta Thunberg–inspired Fridays for Future global climate movement. Why? In part, because while the threat of catastrophic climate change seems far away to many of us, in terms of both time ("it won't happen for several more decades") and distance ("it's not really happening here, it's only happening over there"), young people appear to have a very different view of it.

In other words, the future of the planet has a proximity problem. It's a problem of perception, and if we don't address it, the "far away" will suddenly seem very close indeed. The difference between a global average temperature rise of 1.5 or 2 degrees Celsius will stop being a political talking point, an abstraction, and start having palpable consequences. The number of "climate refugees" will keep increasing, as will the ferocity of their efforts to make it to the "safe harbors" of the Global North, with all of the attendant societal and political consequences.

The argument running through this chapter has been that we need to find better ways to relate to one another, to our shared predicament. Despite emphasizing moral improvements over absolutes such as human rights, this argument is perhaps the most humanist in this book: An expansion in proximity calls on us to design better ways to understand and relate to one another, to shake off our blinders and broaden our horizons—and to do so before it is too late. In the next chapter, we expand this idea and examine ways in which we might stretch our thinking about life itself.

4
EXPANSION 3—LIFE

We all know interspecies romance is weird.
Tim Burton

The clearest way into the Universe is through a forest wilderness.
John Muir

Thirty years ago, the writer Bill McKibben wrote *The End of Nature*, a book widely seen as the first popular account of climate change and our ecological crisis. In it, he warned that a new epoch had dawned. "We have changed the atmosphere and thus we are changing the weather," McKibben wrote. "By changing the weather, we make every spot on the earth man-made and artificial. We have deprived nature of its independence, and this is fatal to its meaning. Nature's independence is its meaning—without it there is nothing but us."

Since the publication of *The End of Nature*, we have for the most part come to assume that we are living in the anthropocene— a subdivision of geological time dating from the beginning of anthropogenic climate change and significant human impact on Earth's ecosystems and geology. (The anthropocene has not been officially declared.) We have now covered, destroyed, or altered half of Earth's surface and dammed a third of its rivers. We use most of the planet's fresh water faster than it can be replenished. And a million species are close to extinction. In a sense, we are designing ourselves out of existence.

This ecological calamity provides the backdrop of the third expansion in this book, which concerns life itself. Indeed, this chapter explores how, even as we alter the planet, probably irreversibly, we are rapidly rethinking what life means.

We'll see how a range of technologies that blur the boundaries between natural and artificial are now within our grasp, from self-programming artificial intelligence to the rapid prototyping of artificial body parts to the production of building materials out of regenerative materials such as mushroom roots.

We'll show how life is changing from something that has always been fixed to something that is open-ended, allowing us not only to extend life but—through the growth of digital afterlife—to "defeat" death. (Up to a point: it is unlikely that we will ever be able to cheat death or become "posthuman," assuming we would even want to.) We'll also see how life is expanding into other dimensions, like virtual reality, the mysterious code of AI-programmed robots, and nanotechnology.

As exciting as these developments are, they come amid growing awareness of our own parlous state of existence, whether from climate change and the increasing inhabitability of parts of the planet, or from deadly pandemics, or even from artificial intelligence. (A couple of years ago, a group of scientists at OpenAI, a nonprofit research company supported by Elon Musk, developed an advanced AI that they deemed too dangerous to be released.)

But as bleak as that sounds, it provides an opportunity for an expansion in our thinking: as the dominance of human-centered design declines, it is steadily being replaced by what we could call species- or life-centered design—a burgeoning movement that seeks to design for the entire planet and takes into consideration other species besides our own. In fact, we might even want to expand our understanding of what "alive" means.

EXPANDING OUR LIFE SPANS

"Who wants to live forever?" Freddie Mercury once sang, answering his own question with the harmonious rejoinder: "This world has only one sweet moment set aside for us." Indeed, for centuries, poets and philosophers have contemplated the eternal conundrum: If you could live forever, would you even want to? Yet recent advances in science and computing mean that it is increasingly less of a thought experiment, more a serious dilemma to address—and one that requires us to consider the implications of extending our lives beyond the limits of what we've thought possible.

The opportunities for innovators will be immensely significant as we rethink concepts such as youth, relationships, aging, learning, and longevity, and extend the human life span, at first through better health care and then potentially though human biological enhancement including gene editing, rapid prototyping of artificial or replacement body parts, brain-computer interfaces, and cloning.

For now, much of this is still very speculative. In an intelligent piece in the *New Yorker* magazine in May 2019, the writer Adam Gopnik determined that living forever remains a pipe dream—and one that, given what it would entail, isn't especially desirable. "What we want," he concluded, "is not eternal life but eternal youth, and what the new science seems to promise us is more like permanent middle age . . . Right now, we live well, and then we don't live well, and then we die. The most that science seems to offer us is this: We'll live well, and then we'll die."

Indeed, the aging of a huge cohort of people—for example, every day between now and 2030, ten thousand baby boomers in the United States alone will turn sixty-five—throws the spotlight back on reimagining death. How can we design for a better death, and better end-of-life care? And how ambitious can we be about designing better and more meaningful circumstances for how we die?

The Re.Designing Death movement, for example, is an informal group of people united by a shared vision: "To create actual societal change on the subject of death. We will never be able to make death a good thing, but we may be able to make it less bad by instigating projects that challenge the taboo, while connecting and celebrating change makers." Among its recent ideas? Wallet-sized "funeral decree cards," designed to be given to partners and family members to help them make executive decisions at the end of their loved ones' lives.

EXPANDING REAL LIFE INTO DIGITAL LIFE

For thousands of years, of course, death meant death. After we shuffled off this mortal coil, we lived on genetically through our children and other descendants, in the memories of those who knew us, and in any artifacts or possessions we left behind.

The advent of digital technology has allowed an even greater sense of personhood to outlast our physical demises. For instance, when we die today, we may live on through social media—at least to whatever extent we chose to build a profile on Facebook, Twitter, Instagram, and so on. (Of course, when those platforms fade into obscurity, who knows what will replace them and exist as digital extensions of our corporeal life on Earth?)

The idea of creating digital versions of ourselves—and of having a digital afterlife—isn't science fiction. Last year, the *Verge* reported the unusual case of Roman Mazurenko, whose soul mate, Eugenia Kuyda, memorialized him as a "chatbot" after he was struck down by a car and killed shortly before he turned thirty-three. Kuyda asked Mazurenko's friends and family to share old text messages and fed them into a neural network built by developers at Replika, her AI startup.

"I didn't expect it to be as impactful," she told the *Guardian*. "Usually I find showing emotions and thinking about grief really hard so I was mostly trying to avoid it. Talking to Roman's avatar was facing those demons."

Similarly, Eterni.me, founded by MIT fellow Marius Ursache, creates a digital version of someone based on their posts and interactions on social media, and will in theory allow people to interact with their loved ones after they've died. Likewise, a Portuguese software developer called Henrique Jorge has created Eter9, a social network that uses AI to learn from the account holder's digital footprint and creates a virtual self, or "counterpart," that lives on after the account holder's death.

A Russian billionaire named Dmitry Itskov wants to go even further—and upload his brain to a computer. According to the *Independent*, Itskov is the founder of the 2045 Initiative, which is working with scientists to develop "cybernetic immortality" within the next few decades. "By perfecting the mapping of the human brain and transferring his consciousness into a computer," the newspaper reported, "he could 'live' much longer—either in the computer, transplanted into a humanoid robot body, or as a hologram."

All of these examples indicate that we may soon, or at least increasingly, find ourselves living in a world populated not only by physically living humans, but by digital ones as well. We could have dinner with the digital edition of our long-deceased grandmother and continue discussing how creamy a perfect mushroom soup should be. What this would mean for family life (besides giving the term *extended family* a new meaning) is one thing. What it would mean for the future workplace is another. Workers in knowledge-intensive jobs need never retire but could continue working via their digital twin, contributing with expertise and organizational history way beyond their own life spans. The extension of bodily life into the digital realm opens a host of opportunities—and dilemmas. (Imagine a company mostly staffed with "cybernetic immortals"—an actual zombie firm!)

EXPANDING LIVING MATERIALS INTO NEW CONTEXTS

Life is also being expanded by pioneers in the realm of the built environment, who are starting to construct what are known as regenerative buildings using renewable and biodegradable building materials.

For example, scientists in the Netherlands have invented a bioconcrete—a concrete that heals itself using bacteria. The bioconcrete is poured like regular concrete but includes capsules of biodegradable plastic containing bacillus bacteria and a food source, calcium lactate. The capsules remain "dormant" for years but dissolve if the concrete cracks and water gets in. That activates the bacteria, which feed on the calcium lactate to produce calcium carbonate—aka limestone—which fills in the cracks.

"It is combining nature with construction materials," explains Henk Jonkers, of the Delft University of Technology. "Nature is supplying us a lot of functionality for free—in this case, limestone-producing bacteria. If we can implement it in materials, we can really benefit from it, so I think it's a really nice example of tying nature and the built environments together in one new concept."

There is similar thinking behind proposals to use mycelium—the network of rootlike fibers that mushrooms and other fungi grow underground. When dried, mycelium can be used to form a super-strong building material that is water-, mold-, and fire-resistant.

For example, a spinout from engineering firm Søren Jensen, in partnership with researchers from DTU, Denmark's national technology university, is publishing a "cookbook" that explains how to grow building materials out of mycelium. With funding from philanthropist group Realdania, the startup is leading a consortium of more than sixteen partners globally to make its recipes available to architects and construction firms. In addition to its properties as a building material, mycelium is biodegradable and so reduces the environmental footprint.

Meanwhile, the California-based startup Mycoworks uses mycelium "to create sustainable alternatives to leather, plastic foams, and other problematic materials." The company says it has "created a new kind of leather grown rapidly from mycelium and agricultural byproducts in a carbon-negative process." Indeed, its custom-engineered material is sustainable, versatile, and animal-free, and looks and feels like leather. Similarly, Ecovative is a materials science company that's developing "a new class of home-compostable bioplastics" based on mycelium. It dubs them "Mushroom Materials" and describes them as "high-performance, environmentally responsible alternatives to traditional plastic foam packaging, insulation, and other synthetic materials."

These new innovations in materials expand our perspective of what is alive and what is not—and open new avenues not only for creating more sustainable buildings but also deploying nature's own resources in shaping everyday products that affect all aspects of our lives. As we will discuss in the next section, this is only the beginning of a revolution that allows us to manipulate all living things in new ways.

EXPANDING LIFE THROUGH LIVING MATERIALS

In his best-selling book *Homo Deus*, Yuval Noah Harari writes, "You may not agree with the idea that organisms are algorithms, and that giraffes, tomatoes and human beings are just different methods for processing data. But you should know that this is current scientific dogma, and that it is changing our world beyond recognition."

That a preconception of what biological life represents is shaping our future is not evident to most people. The technology that arises as a consequence will strengthen that worldview. If we want technological diversity, we need to have an open discussion about what biases certain technologies have—and, in general, the philosophy of technology and science.

Synthetic biology—which allows us to write and edit genes—typifies this shift. Indeed, our newfound ability to engineer life on

Earth is nothing more than a fundamental change in the way we interact with it. As the *Economist* put it, "It permits the manufacture of all manner of things which used to be hard, even impossible, to make: pharmaceuticals, fuels, fabrics, foods and fragrances can all be built molecule by molecule."

One of the arguments of this book is that if we are going to have any chance of addressing long-term goals like the UN's Sustainable Development Goals (SDGs), we need a different level of thinking. In that context, we aim to provide a series of expansions that can be used as a resource for any individual or organization interested in addressing the SDGs and the goals that will succeed them. Indeed, many of the innovative new solutions espousing expansive ways of thinking are spurred by the ambition—by states, companies, or individuals—of addressing the SDGs.

The bioscience company Chr. Hansen offers a very good example of how life-forms can be used to address SDGs. The 140-year-old company "develops natural solutions for the food, nutritional, pharmaceutical and agricultural industries," and was named the most sustainable corporation of 2019 by *Corporate Knights*. According to the Canadian magazine, Chr. Hansen's sustainability approach is based around how it can contribute to the SDGs, with 82 percent of its revenues contributing to the targets.

One focus, for example, is on bacteria that can reduce food waste by prolonging the shelf life of food products such as yogurt and cheese. Chr. Hansen is aiming to reduce yogurt waste by 1.2 million metric tons by 2022 from 2015 levels. Another focus is on sustainable agriculture. "We provide natural microbial solutions that can be used as an alternative to conventional pesticides while also increasing yields," the company's CEO Mauricio Graber told *Corporate Knights*. "It is much closer to the way that nature intended things and removes the risks and negative impacts of pesticides."

Chr. Hansen is also looking to address the SDG of health and well-being and to try to reduce the overuse of antibiotics in livestock farming and the impact of antibiotic resistance through probiotic bacteria. According to *Corporate Knights*, Chr. Hansen's probiotics products "help infants to develop good gut health; others can shorten the duration of the common cold and influenza-like illnesses."

Chr. Hansen's work with bacteria represents an expansion from working with visible to invisible life, from small to ultra-small. In the same vein, other visionary individuals and organizations see potential in algae. Yes, algae—as in that icky green goo that clogs waterways and covers the surface of lakes and ponds. Like trees, you see, algae use photosynthesis to convert carbon dioxide and water in the presence of sunlight into sugars and other nutrients. In industrial settings, this could be a game changer. Expanding our concept of building blocks can literally expand and extend life.

Consider the Algoland project, at the Degerhamn cement factory in Sweden. Environmental scientist Catherine Legrand and a team of researchers at Linnaeus University have found a way to get naturally occurring algae to capture carbon dioxide coming from the factory before it enters the atmosphere—and convert it into oxygen. According to *Quartz*, by running the factory's noxious emissions through the photosynthesis system a few times, the algae will remove almost all of the greenhouse gas. And that's not all: the algae are rich in protein and fat, and after drying, it can potentially be used as an additive for chicken and fish food.

All of which begs the question: Where might this expansion end—if at all? Does rethinking life have any real limits? We should remember the idea that digital technology, based on electrons, is arbitrary—that, in theory, an alternative technology based on chemicals could substitute it. Which is not that far-fetched. Consider the notion of the wetware computer—the computer that is made of organic material such as living neurons, and which is therefore thought to be capable of "thinking for itself." Wetware is conceptual—for now. But it's an example of how our thinking needs to be different and how we need to expand our understanding of what's living. Arguably, it's even evidence of how our digital age wasn't predetermined—that decisions directed discoveries and developments that led to our digital age. In other words, it wasn't set in stone that we would have Motorola StarTACs and other digital devices.

EXPANDING THE SCALE OF ARTIFICIAL LIFE

"The robots are coming" is an all too familiar headline these days. The story tends to be equally predictable. From the end-of-days forecast of robots destroying humanity (with the plot ripped from the Terminator movies and all robots equated with Skynet) to the comparatively benign fears of robots taking over our jobs, there's no shortage of alarm. Less attention is paid to the potential benefits of robots—or, as a recent V&A publication *The Future Starts Here* wondered, how a world of robots might challenge our understanding of what it means to be human.

Like many of the newfangled ideas explored in this book, tomorrow's robots will vary in scale, from bacteria-size robots used by doctors to unclog arteries or treat diseases, to mosquito-size drones used by the military and intelligence agencies, to domestic robots employed in the home, to self-driving vehicles, to giant, industrial ones used in factories or in the planet's less hospitable regions.

For now, robots seem to have the greatest application in the labor market, doing everything from assembling cars to picking fruit. Job losses are inevitable in the short term—but new spheres of employment will emerge. And the paradigm doesn't have to be binary. Humans and robots will increasingly act in new and more sophisticated constellations. When not taking part in chess tournaments, human-robot teams could collaborate in other ways (something we'll explore further in chapter 6).

To take an example from science fiction: in the movie *Pacific Rim* (2013), we follow the unlikely hero Raleigh and his brother Yancy. "You wouldn't have picked my brother Yancy and I for heroes," Raleigh tells us. "No chance. We were never star athletes, never at the head of the class, but we could hold our own in a fight. And it turned out we had a unique skill. We were drift compatible." Being "drift compatible" means the brothers are neurally linked and able to work as a team to control a giant robot, the Gipsy Danger. The brothers are neither the brightest nor the strongest, but they make a great team. And compared to other robots, Gipsy Danger is neither the strongest nor the best equipped. But it is the most successful, thanks to its formidable partnership with the two brothers.

In today's world, the startup Universal Robots is pioneering the use of so-called cobots, which are designed to share a workspace with humans and make automation easier for businesses by giving them "access to all the benefits of advanced robotic automation, without the extra costs associated with traditional robots," such as difficult programming or long setups. Cobots are lightweight, easy to redeploy, and can be given "dangerous and dull" jobs in order to reduce repetitive strain and injuries to humans.

EXPANSION 3—LIFE

While many robots will simply replicate human tasks and carry them out more efficiently, might others transcend their role to cease being "just" an algorithmic, AI-programmed machine but instead be seen to be "living" alongside us? The failure of so-called "social robots," such as Jibo, Kuri, and Anki, suggests that it is hard to make appealing domestic robots that aren't drones or vacuum cleaners. On the other hand, the "advanced interactive robot" PARO has been used in Japan and Europe since 2003. Resembling a baby seal, it blinks and coos when petted, and is therapeutic for patients with dementia.

Of course, the expansion in what we understand to be living—or at least the blurring of the natural and the artificial—is not limited to robotics. For example, we are getting closer to making laboratory-grown meat available for commercial use. Lab-grown meat—aka clean or synthetic meat—is meat produced from animal cells grown in a laboratory rather than from slaughtered animals. In theory, it requires fewer natural resources to make, results in fewer greenhouse gas emissions, and ends animal rights abuses.

Although few companies have yet announced plans to make their food available for commercial use, much less for public consumption, startups such as Memphis Meats, JUST, and Finless Foods are all hoping to grow many meals' worth of meat from a handful of cells. "Theoretically from one little piece of meat you can create an unlimited amount," Mike Selden, CEO of Finless Foods, told *Wired* last year. A milestone moment for the industry came in December 2020 when "chicken bites," produced by the US company Eat Just, passed a safety review by the Singapore Food Agency. The approval presages a future in which all types of meat can be produced without killing livestock.

What this technological shift means is that while computing has expanded from rudimentary machines to wearable interfaces to artificially intelligent systems that are increasingly lifelike, farming is becoming increasingly distanced from the natural world. For thousands of years, humans hunted animals for food and tended livestock in small farms. Then, in the last century, we industrialized food production and turned farms into factories. And now, in the first quarter of the twenty-first century, we are poised to turn the lab into the food factory and potentially eliminate the need to farm animals at all.

The way we're going, it is not that outlandish to suggest that we may one day employ AI-powered "robo-chefs" that are more "alive" than the lab-grown beef or fish they cook for us. The dystopian endgame, of course, would involve the robots rising up to dominate humans, eventually reducing *us* all, Matrix-style, to lab-grown entities. A less alarming version of this inversion was captured by the *New Yorker* in May 2019. Titled "Dog Walking 2.0," the cartoon on the cover depicted a human taking a robot dog for a walk and crossing paths with a robot with its own *flesh-and-blood* dog on a leash. The two dogs, real and robot, are giving each other a look, perhaps in recognition of the situation's uncanniness.

What these examples show us is an expansive view of what it means to be alive. We see the emergence of collaborative and social robots, which means we as humans will increasingly relate to technological innovations as living organisms. At the same time, we are replacing living organisms as sources of nourishment, making what was once the most basic and simple human activities—such as making meals from what nature offers—into a product of advanced engineering. Things are getting flipped on their heads.

EXPANSION 3—LIFE

EXPANDING FROM HUMAN- TO PLANET-CENTERED DESIGN

The Swedish statistician Hans Rosling wrote a much-admired book called *Factfulness* that explains how we're wrong about the state of the world and how things are better than we think: steadily falling rates of world hunger, child mortality, HIV infections, numbers of nuclear warheads, deaths from disaster, and so on. Yet the natural world gets only a handful of mentions—in a pair of graphs showing the increase in Earth's land mass that is protected and in the growing number of protected species. In other words, the world is apparently getting better—but seemingly only for us. Put any other animal, plant, or even the planet per se in focus and things are not improving. Rosling wrote a necessary book, but the absence of nature is glaring.

"As humans we've been a little bit silly," says the Zimbabwean-born designer Natsai Audrey Chieza, who works at the intersection of creative and biotechnology industries. "We've been designing for one species for a really long time. The question is, how do we move away from human-centered design to a multispecies agenda for design? Which means that when we're designing, we aren't just thinking about me, me, me. We're thinking about us—both living and nonliving, human and nonhuman."

Chieza recognizes it's provocative to say we need to change our entire systems of thinking and think "beyond us," and that a significant cultural mindset shift is required. One way of achieving it might be to get people to understand the symbiosis between different organisms and to "engage with the hidden lives, if you like, of these organisms." And that is slowly starting to happen as we discover more and more about the natural world.

EXPAND

"We need to reorganize who is above whom," argues Olafur Eliasson. "We need to horizontalize the notion of power. Some people are now giving trees the rights of personhood, which is very interesting. Economically speaking, who owns the wood of the tree? Maybe the tree. As a design exercise, I think we should ask ourselves, with what right can we choose a tree for a table?"

The Hidden Life of Trees, the best-selling book by German forester Peter Wohlleben, shows us that trees are social beings that communicate with and nurture each other, share nutrients, and contribute to a broader ecosystem. Similarly, the ecologist Suzanne Simard discovered "a vast tangled web of hair-like mushroom roots—an information superhighway allowing trees to communicate important messages to other members of their species and related species, such that the forest behaves as 'a single organism.'"

Consider, too, farmers such as Allan Savory, who argue that the "holistic planned grazing" of cattle can help stop desertification and climate change. Though his claims and methodologies are highly controversial—and disputed by scientists—they show that even as our manipulation of life expands into the domain of science, there are areas in which it could expand outside the laboratory—in a way that puts the planet and other species first for a change. In any case, our more informed understanding of how the natural world works requires us to reconsider what we deem worthy of protecting—and designing for.

This raises an interesting question: What is nature worth? Cambridge academic Sir Partha Dasgupta has produced a comprehensive review for the British government titled "The Economics of Biodiversity," which essentially seeks to determine what value nature (or Nature with a capital "N," as Dasgupta purposefully spells it) represents for society. He concludes that our entire economies, livelihoods, and well-being all depend on "our most precious asset: Nature." The report asserts that as a species, we have engaged unsustainably with nature and that this is endangering the prosperity of current and future generations. And then the kicker, which truly calls for expanding our perspective and rethinking our actions: "At the heart of the problem lies deep-rooted, widespread institutional failure." This failure is the more critical because our economy, according to Dasgupta, is not external to nature, but fundamentally embedded in it. So we need to treat nature as such.

Fortunately, some institutions are assuming the responsibility of rethinking our engagement with the natural world. Perhaps it is not surprising that design schools are among those at the forefront. For instance, at the design school in the Danish city of Kolding, the Design for Planet program "views the value of design as a means of generating changes in the way we do things" and both directly and indirectly addresses the UN SDGs. According to the school, "Design for Planet encourages free and critical thinking in relation to the sustainable agendas it addresses and the program embraces an understanding of the designer as a transformer of what exists here and now as well as a generator of novel solutions so that design can contribute as much as possible to a more sustainable world."

The expansive thinking in play here is the ability to take a much wider perspective of life: from life as a uniquely human reality to a reality that embraces everything living (in the widest of senses) as the stuff we design with, and for. The question, however, becomes: What happens when our creations take on a life entirely of their own? We'll examine this further expansion of life in the next section.

EXPAND

EXPANDING FROM HUMANITY TO POSTHUMANITY

The move away from human-centered design is needed, but it still assumes that humans will always be part of the equation. At our current rate of "progress," we might not want to take that for granted. Acolytes of philosopher Ray Kurzweil positively welcome the Singularity—the moment that artificial intelligence surpasses human intelligence—even if the outcome isn't guaranteed to be as clear-cut or beneficial for us.

We can't say the warning signs weren't there—and not just in provocative feature films such as *Transcendence* or the Terminator series, none of which point to a rosy future for our species once a more intelligent being comes along. For example, researchers at Facebook became alarmed a while back when they discovered that some of their chatbots were communicating in a new language—or what *Fast Company* later described as potentially "the most sophisticated negotiation software on the planet." And while Facebook didn't exactly pull the plug on it, it did ultimately elect "to require its negotiation bots to speak in plain old English."

This story sounds alarming, but it really shouldn't be, because the truth is we don't always understand what AI is doing. "Artificial intelligence is making decisions by reviewing people's medical tests in hospitals, credit histories in banking, job applications in some HR systems, even criminal risk factors in the justice system," explained *Fast Company*. "Yet it's not always clear how the computers are thinking."

Adam Greenfield, author of *Radical Technologies: The Design of Everyday Life*, makes a similar argument—but goes further. "It's not just that we don't understand the technologies we use," he argues. "It's that we *can't* understand them." In particular, he cites blockchain as an example of an emerging technology that is not well understood, even by its creators. He calls for more "social design"—which he says is increasingly important in a world where we are codifying almost every aspect of human life.

Silicon Valley determinists widely assume that machine consciousness is merely an incremental computational capacity analogous to their vernacular technologies. But this is simply an unknown still. It is accepted that mental activity in some way correlates to the behavior of the material brain. However, alternative theories are not yet extensively explored. Quantum theory, for instance, is the most fundamental theory of matter we have. It introduced an element of randomness standing out against the deterministic worldview, and it would be legitimate to ask whether the theory can help us understand consciousness.

Machine consciousness may only come if we want it to—and start understanding the consequences of it, which we can only speculate about. Exhibitions such as *Broken Nature*, at the 22nd Milan Triennale, provocatively raise the specter of a posthuman world. "Even to those who believe that the human species is inevitably going to become extinct at some point in the future, design presents the means to plan a more elegant ending," explains its curator Paola Antonelli. "It can ensure that the next dominant species will remember us with a modicum of respect: as dignified and caring, if not intelligent, beings."

In Alex Garland's 2014 science fiction thriller *Ex Machina*, a mysterious CEO who has created a female humanoid robot with AI offers a chilling prophecy to a visiting computer programmer. "One day, the AIs are going to look back on us the same way we look at fossil skeletons on the plains of Africa," he says. "An upright ape living in dust with crude language and tools, all set for extinction."

EXPAND

Thirty years after he wrote *The End of Nature*, Bill McKibben revisited the theme of climate catastrophe and species extinction in *Falter: Has the Human Game Begun to Play Itself Out?* The celebrated environmentalist sets the tone in the opening chapter. "I think we're uniquely ill prepared to cope with the emerging challenges. So far, we're *not* coping with them." But, he adds, "there is one sense in which I am less grim than in my younger days. This book ends with the conviction that resistance to these dangers is at least possible."

Indeed, McKibben eventually finds grounds for optimism in the power of nonviolent movements and in the capacity of solar panels to provide billions of people around the world with cheap renewable energy. In the end, though, it all comes down to will. "We can wreck the Earth as we've known it, killing vast numbers of ourselves and wiping out entire swaths of other life," he concludes. "But we can also *not* do that."

We agree that we have it within ourselves to ensure that this fossilized future isn't our destiny. Whether we interpret our gravest threat as climate change, or species extinction, or machine consciousness, it remains within our grasp to design our way out of the mire, to expand our thinking, and to preserve the greatest gift that we have ever received: that of life itself and all that it entails. In the next chapter, we consider ways to expand our understanding of value—and, in doing so, improve our stewardship of the most valuable thing we know. Our planet.

5
EXPANSION 4—VALUE

> We can make all of humanity successful through science's world-engulfing industrial evolution provided that we are not so foolish as to continue to exhaust in a split second of astronomical history the orderly energy savings of billions of years' energy conservation aboard our Spaceship Earth.
> R. Buckminster Fuller, *Operating Manual for Spaceship Earth*

> At its core, the new economy is based on culture: on the culture of innovation, on the culture of risk, and the culture of expectations, and, ultimately on the culture of hope in the future.
> Manuel Castells Oliván, *The Internet Galaxy*

A couple of years ago, a young Taiwanese engineer named Arthur Huang was standing on the rooftop of a building in Taipei, enjoying a cigarette break with three of his colleagues at Miniwiz, the design firm he founded in 2005. The conversation turned to the cigarettes they were holding.

"We were like, 'What's this made from?'" Huang recalls today. "So we started peeling open the paper and tearing them apart."

EXPAND

For all the pleasure it provides and the damage it does, a cigarette is a simple thing. Typically, between 80 and 120 millimeters in length and eight in diameter, a standard cigarette weighs just over a gram—the majority of it tobacco. Then there's the filter, the centimeter-long squidgy tip that's meant to make smoking seem less unhealthy. "We were like, 'Dude, this is a sponge. What material is it?'" says Miniwiz's forty-year-old CEO. "We started looking into it, and that led to a three-year journey into material innovation."

To understand why Huang and his colleagues got so interested in filters, consider this: cigarette butts are believed to be the most common form of man-made litter, with 4.5 trillion dropped every year. Some end up in sewers and water purification plants, others in the sea. Indeed, plastic straws aren't the biggest ocean contaminants; cigarette butts are.

Which is precisely why Miniwiz's designers and engineers began to look into them. The company's mission is to explore the unlimited potential of rubbish by taking recycled materials to the highest levels of product engineering. In partnership with Nike, for example, Miniwiz produced packaging out of milk cartons and coffee cup lids. It also constructed a chair out of spent beer grain, and a nine-story pavilion using 1.5 million recycled plastic bottles. The intention, Huang often likes to say, is to "make trash sexy."

Miniwiz soon discovered that cigarette filters are commonly made of cellulose acetate—which really piqued their interest. "Cellulose acetate is used in gauze, it's used in air filters, and it's used in sunglasses," Huang explains. "What an interesting application—cigarettes and sunglasses. The perfect link."

In time, Miniwiz developed a process for separating cellulose acetate from butts collected from bins outside tobacco giant Philip

EXPANSION 4—VALUE

Morris's offices in Switzerland and Italy. Then it compressed the cellulose acetate into boards, creating the base material for manufacturing sunglasses and buttons—an item historically made from cellulose acetate. Huang hopes it will one day be broadly used as a yarn, in 3-D printing, and in construction too.

Miniwiz is one of a handful of companies seeking to bring hard-to-recycle materials back into the economy and extract value out of the planet's growing pile of rubbish. "We're trying to turn pollution into a solution," says Huang. In an ideal world, there'd be no pollution at all—or at least the most sustainable solution would be to prevent pollution rather than have to deal with it. But no one's figured that out. "This unsurmountable pile of trash that we're constantly generating is just being misplaced, displaced, and not utilized," adds Huang. "And that isn't just an engineering challenge—it's a huge business opportunity."

Let's not lose sight of the fact that cigarettes are also deadly and addictive; in an ideal world, nobody would smoke. Until that day comes, however, we have to face the reality that cigarette butts are a nuisance and a trash problem. Doubtless, Miniwiz would prefer not to have to collect cigarette butts from recycling bins, but until it no longer needs to, it makes sense for the company to return them to the economy and extract value from it.

In any case, what Huang calls "urban mining" illustrates the kind of innovative thinking the world needs right now—and it embodies the expansion that we discuss in this chapter, which is to say the expansion of value, and in particular how we extract, manufacture, trade, own, consume, and discard our planet's resources.

In this chapter, we make the case for further shifting our thinking away from the old, linear model of producing and consuming resources and instead adopting circular, networked, and multidimensional models of design, use, redesign, and reuse. We argue that we ought to introduce additional dimensions of value, beyond profits and shareholder value. And we make the case for expanding our view of value creation, to see it not as a chain or even a circle, but as a networked system, where multiple actors interact and collaborate. But first we take a step back and try to get a handle on the problem facing us all.

EXPANDING IN THE ANTHROPOCENE

"This is the largest clarion bell from the science community, and I hope it mobilizes people and dents the mood of complacency." Debra Roberts was in no mood for equivocation. As cochair of the UN's Intergovernmental Panel on Climate Change working group on impacts, she and other leading climate scientists had just warned that humankind has only a dozen years to limit catastrophic climate change.

According to the IPCC, global warming must be kept to a maximum of 1.5 degrees Celsius, beyond which a rise of even half a degree will significantly aggravate the risks of drought, floods, extreme heat, and poverty for hundreds of millions of people, as well as loss of biodiversity amid increasing habitat loss and land degradation. "It's a line in the sand, and what it says to our species is that this is the moment and we must act now," Roberts added.

Unofficially, we are now living in the anthropocene—an epoch defined by human-influenced climate change and our significant impact on the planet's geology and ecosystems. Earth's resources are running out. Our industrial system is reaching its physical limits. The linear economy has wrought environmental havoc. To stop it, we need to think differently about every aspect of our lives, from the food we eat to the clothes we wear to the products we use. And these requirements are getting more urgent. The world's middle class is growing at an unprecedented rate. According to the Brookings Institute, at a global level, we are witnessing the most rapid expansion of the middle class the world has ever seen. At the end of 2016, there were about 3.2 billion people in the global middle class, with 160 million joining it every year, on average, for the next five years.

The consequences of this shift will be unprecedented too. "The changing distribution of middle-class spending toward new entrants will have an effect on markets," explained Homi Kharas of the Brookings Institute. "Households just entering the middle class will seek to purchase consumer durables as well as services, including tourism, entertainment, health, education, and transport." More to the point, as the income of the global middle class grows, so too will their carbon footprint.

In response to the looming disaster, a sea change in economic thinking is taking place. New business models are being proposed almost daily. For example, in *The Zero Marginal Cost Society*, economist Jeremy Rifkin argues that we are entering a world beyond markets and learning how to live together in an increasingly interdependent, global collaborative commons. And Kate Raworth, in *Doughnut Economics*, says we've had a century of chasing the false goal of "endless growth," which has pushed us to the point of ecological collapse. She calls for a new goal—to meet everyone's needs within the planet's means, or what she calls the "sweet spot for humanity." To have half a chance of getting there, she wonders, "What economic mindset would be fit for the task?"

EXPANDING THE CIRCULAR ECONOMY

One answer is to stretch our thinking about our "take, make, dispose" extractive industrial model and adopt a circular model instead. According to the Ellen MacArthur Foundation, which works with businesses, governments, and academia to try to accelerate the transition to a circular economy, the model "entails gradually decoupling economic activity from the consumption of finite resources and designing waste out of the system."

The concept of the circular economy is based on three principles: "to design out waste and pollution; keep products and materials in use; and regenerate natural systems." It's about giving a second (and third, fourth, fifth, etc.) life to materials, through repairs or upgrades, or by leveraging technology to recycle and reuse materials.

The point isn't simply to mitigate the negative effects of the extractive linear model. It's to create long-term economic and environmental resilience, to benefit both people and the planet, and to create new business opportunities. Indeed, the circular economy isn't just about hugging trees. Accenture has estimated that circular economy models could generate as much as $4.5 trillion US in additional economic output as early as 2030.

"The driver of the circular economy isn't scarcity—it's opportunity," says Peter Lacy, managing director for growth, strategy, and sustainability at Accenture. "By keeping resources economically productive for as long as possible, companies can achieve greater growth. Most companies have waste hot-wired into their existing ways of doing business, and it will take many steps for most to turn waste into wealth. But those who get there first will achieve circular advantage that differentiates them in their market." Little wonder, then, that many major brands, including

EXPANSION 4—VALUE

Nike, Adidas, and H&M, have started to explore circular models. Adidas, for instance, has developed a pair of sneakers made using plastic waste recovered from the ocean.

Then there's Carlsberg. "Every generation has a responsibility to give the world back, at least in the same condition as we received it from our parents," Flemming Besenbacher, former chairman of the Carlsberg Group, told us. It's a philosophy that explains why the former scientist believes there isn't just an ethical case to rethink how the Danish brewer does things, but a business case. Consumers will increasingly evaluate a company's impact on the world, he reckons, and refuse to buy products from companies that aren't sustainable.

"Brewing for a better today and tomorrow" is Carlsberg's stated purpose now, and it has implemented ambitious plans to decimate its carbon footprint as well as the amount of water it uses. That means investing in long-term scientific research—such as developing new strains of yeast and barley—and rethinking everything from bottling to packaging to distribution. A case in point is the company's new style of six-pack: the cans are held together with glue instead of planet-polluting plastic.

Then there's circularity. "When you design any product today, you should be thinking circular," Besenbacher says. "It might not be possible for the product you're designing right now, but you should have this circular mindset in whatever you're doing—to reduce, reuse, recycle, and rethink as much as possible."

It's no exaggeration to say that Carlsberg is driving circular policy in Denmark and beyond. Indeed, Besenbacher duly received the former 2019 Fortune Award for Circular Economy Leadership. He believes we need to see a paradigm shift in production and design. "When you are already constructing your product, you should immediately be thinking about how it can be dismantled, taken apart, how some of it can be reused, how some of it should perhaps be destroyed in a good manner, and so forth."

In 2015, Besenbacher unveiled a prototype of a zero-waste beer bottle at the World Economic Forum in Davos. Developed in partnership with design firm Kilo, the bottle is made from sustainably sourced wood fibers—meaning that, unlike bottles and cans, which can be recycled but don't degrade if improperly discarded, it is 100 percent biodegradable. "We are constantly trying to innovate, to develop new ideas, and see how they can be embedded into our business case," says Besenbacher.

Of course, some may say that the circular economy is simply a case of old wine in new bottles—a fancy word for recycling. Creative reuse of materials is certainly a part of it. But it's wrong to think of circularity as *solely* about turning chip bags into tote bags, say, or fishing nets into wash bags. What makes the circular economy so interesting is how it *creates value*. Essentially, it decouples growth from the use of new resources and materials by extending the life cycle of resources in use and keeping them in circulation. Another way of putting it is that if linear business models are about ensuring the *flow* of goods and materials, circular ones are all about maintaining the *stock*.

For years, our understanding of value creation has been largely one-dimensional. A product's life cycle was fixed and linear. Value creation existed in a chain that began with the extraction of a raw—or rare—material from the earth and led to the creation of products. Value creators were individuals and businesses who added different layers to the products as they were refined and distributed toward the point of sale. Value was destroyed by consumers who bought, used, and eventually threw away those products.

EXPANSION 4—VALUE

Today, though, we're seeing a shift in thinking about what constitutes value. One of 2018's most celebrated books was *The Value of Everything: Making and Taking in the Global Economy* by Mariana Mazzucato, founder and director of University College London's Institute for Innovation and Public Purpose. Arguing that modern economies have shifted from creating value for the benefit of all to rewarding activities that extract value for asset owners, Mazzucato advocates a "mission-oriented stakeholder model of capitalism where value is created collectively." (She also believes governments should become value creators rather than mere facilitators or spenders. We argue too, in particular in chapter 7, that we must expand our conception of the role of the public sector in creating value.)

Thanks to the circular economy, business models and concepts of value creation are changing. In the circular economy, products are designed to create value by their very circulation. True circularity is thus about creating a model of supply and manufacturing that allows materials to be used or sourced again and again. Instead of being recycled once, a material can be upcycled or renewed many times over, or disassembled and reused for different purposes.

In a truly circular model, products are also made to last as long as possible, in a way that makes it possible to separate and reuse the different materials and components again. Imagine a running shoe whose constituent parts could be removed and replaced when they had worn out. Instead of buying a new pair, you would simply send back the parts that needed replacing and receive custom-3-D-printed replacements. (And, as you pound the pavement, the old parts of your shoes are recycled into new parts for someone else.)

EXPAND

The circular economy thus requires us to flex different mental muscles and rethink how we design products, systems, services, and organizations. It requires several shifts in perspective—including, for instance, thinking about how to make products that are designed to be made again. Faced with a fast-growing population, dwindling resources, and the threat of catastrophic climate change, this shift in thinking is vital. For designers, it is nothing short of revolutionary. As the Ellen MacArthur Foundation puts it, it provides an opportunity to "redesign the way we make stuff."

We, too, believe design will play a vital role in the expansion of the circular economy and deliver impactful change. Designers are able to think across traditional materials, methods, sectors, and professions—and produce desirable products and services. The circular economy challenges designers to think more expansively, to design products not with single-use or planned obsolescence in mind but with built-in extensions to their life and use. Extensions in how long we use them because of the personal attachment we form to them, or because of the timelessness of their design, or because we can share it with others once we're done, or because we're able to adapt it to our changing needs.

Designers can also extend the life of products by making them with high-quality materials and techniques, or because they can more easily be repaired or replaced, disassembled and reassembled, or made with materials that can be recycled technologically, or that will one day return to the biosphere and be recycled biologically. Designers can help us produce goods that we need—goods that meet challenges—rather than goods that we just want and that cause more social or environmental problems than they are seemingly worth.

Circular design is also about rethinking products and business models from the ground up so that growth and consumption can continue without using new resources or materials—and without compromising either aesthetics or quality. And in that respect, both established brands and startups have plenty to gain from the circular economy.

For businesses, however, it's not about selling as many products as possible. It's about producing the best possible product, one that can be kept in circulation and reused and recycled. The aim for designers, then, is to produce items that are aesthetically appealing, that don't compromise on cost or quality, and that can be used in the circular economy.

Might Scandinavia lead this shift? After all, the region has long been known for its well-designed and reliable furniture: chairs, lamps, and beds that last a lifetime and are passed from one generation to the next. As it happens, a champion of circularity is IKEA. In 2018, it won the Accenture Strategy Award for Circular Economy Multinational at the World Economic Forum in Davos, in recognition of its work leading to sustainable transformation of its business in everything from product development to material sourcing to logistics.

For instance, IKEA found a way to transform used PET bottles into a foil that is laminated onto KUNGSBACKA kitchen fronts—the first range of kitchen fronts made from both recycled wood and recycled plastic. It is also testing a new service that will allow people to make their old IKEA furniture available for resale or reuse. "It's about smarter use of resources and, from the very beginning, designing products so they can be repurposed, repaired, reused, resold or recycled," said IKEA of Sweden's managing director, Peter van der Poel.

Scandinavian design is also renowned for its aesthetic appeal—that is, its simple, functional minimalism (or essentialism). Here's the thing. Aesthetic appeal will be central to the success of the circular economy. Simply put, if people lose their emotional attachment to a product (or if they never develop one at all), they're more likely to lose or discard or destroy that product. Jens Martin Skibsted coined the concept of "aesthetic sustainability" to describe this phenomenon.

Albeit in a different way, aesthetic sustainability certainly seems to underpin Miniwiz's crusade to "make trash sexy." According to founder Arthur Huang, it's crucial that a recycled material meets the same level of quality that people are used to. "A recycled water bottle has to turn into something of quality so that people are willing to look at it and consider it an alternative," he says. Huang's aim is to make new materials out of waste and for people "to not even question whether it's made from recycled material anymore." For example, he wears trousers made of recycled polyester—and which fool people into thinking they're woolen.

By the same token, if consumers are attracted to a product's look and feel—its durability and quality, heritage and craftsmanship—they're more likely to hold on to it. Visit any design museum in the world, and you'll see that the items on display are there because of their long life cycle coupled with their function and aesthetics. Examples include chairs designed by Danes like Hans J. Wegner, Arne Jacobsen, and Børge Mogensen. A case in point, a Danish taxi driver once told us he was rethreading his Wegner chairs so that his son could have them, while one of us owns a Wegner "Bear Chair," now used by the fourth generation. The chairs still retain their functional and aesthetic value after more than half a century and continue to be in demand.

At the geopolitical level, the European Union is also embracing circularity in a big way in its bid to counter climate change. A particularly intriguing initiative is the "New European Bauhaus," ostensibly conceived by European Commission President Ursula von der Leyen. The aim of the program, launched in late 2020, is to invite architects, designers, and artists to lead the green transition of Europe's urban environment, with a particular emphasis on aesthetics, social inclusion, and circularity. Time will tell if this program succeeds in renewing the socially radical spirit of the highly influential original Bauhaus school of 1920s Germany for a twenty-first-century context.

Let's be clear: circularity isn't a silver bullet. Given the scale of the problem, we'll need a number of solutions, one of which is switching from a linear to a circular model. There are hurdles, too, when it comes to implementing circularity. For one thing, while almost everything can be recycled, collecting and separating materials is a challenge, as is the contamination of materials. Moreover, there's only so much reuse and recycling a material can take. Simply put, the recycling process shortens a material's molecular chain: as you shorten the chain, the material gets weaker.

There's also an inherent limit to what we can do with some materials. For example, we don't need more tote bags made from recycled plastic. To put it another way, just because something was reused or recycled, it doesn't mean any value was added.

Consumer behavior also needs to change. Right now, too many of us find it too easy simply to throw materials away rather than recycle or reuse them—or get them to companies such as Miniwiz. And at the other end of the scale, there simply aren't enough companies like Miniwiz yet.

"I feel a little bit lonely," Huang concedes gloomily. Moreover, Miniwiz and others like it, such as TerraCycle—which turns scrap denim into messenger bags, surgical gloves into park benches, and old computers into plant pots—are still dwarfed by the size of the problem.

Of course, many of these challenges are ones that designers can try to solve. But Huang reckons that governments also need to show more leadership. "If the UK banned disposable cups," he says, "just imagine how many new business ideas and new design or engineering companies would come along to try to solve the problem." He points to the recent ban on plastic straws, which triggered a tidal wave of solutions, from bamboo straws to wax paper straws to biodegradable ones. "All of a sudden, it was like the Wild West, and all that is energy," he says. "That's what we should be aiming for."

There seems no doubt that governments have a key role to play in accelerating the changes underway, including through appropriate regulation. The world's leading economies could demonstrate more leadership and incentivize a faster transition to a circular economy. The European Union's "New European Bauhaus" initiative looks promising in that respect. But countries that have shown some leadership—including Scandinavian nations—could still do more.

EXPANDING TO VALUE-CREATING NETWORKS

Just as we need to accept the idea that circularity isn't a silver bullet, so too should we expand the view that circularity is an endpoint. There is, after all, a slew of potential business models and platforms, including circular ones. If we were to expand circularity, what would it look like? What comes after circularity? What might the next paradigm be?

One answer is that having expanded our thinking about economic models to consider circularity as a better alternative to linearity—and one that reconsiders how we create value—we now need to stretch even further and conceive of value creation as something that is created within systems or networks.

Instructive in this respect is the recent thinking of academics Rafael Ramírez and Ulf Mannervik, of Oxford University's Saïd Business School, who build on the work of strategic thinker Richard Normann. From the 1980s to the 2000s, he developed an original approach to strategy that sees value as inherently coproduced in systems—an approach that stood in contrast to the then-dominant notion of competitive advantage.

Ramírez and Mannervik believe businesses can redesign their products or services by changing their roles in the wider value-creating system of which they are a part. In *Strategy for a Networked World,* they make the case for visualizing the wider system in which a firm or organization creates value and for mapping all the relationships and value exchanges in that system.

They also take a broader view of value than a purely financial one. Their conception includes dimensions such as knowledge, trust, and brand. This thus expands our understanding of the contributions that current or new offerings can make to a system, and becomes a way to innovate the offerings provided, or even the positioning of the organization in the system in which it wishes to provide value.

Indeed, a key to the success of the circular model are wider networks and sharing platforms that allow companies to squeeze additional value out of their products. Instead of owning something, we share it and rent it out on an as-needed basis—baby

EXPAND

clothes for as long as they fit our newborn, say, or a car for as long as we need it to drive to the store and back. For example, Furlenco is an Indian online furniture platform that allows customers to lease its products—and swap items whenever they move to a new house or simply get tired of them. What's more, it might become more relevant to ask how much a product costs per year of ownership *regardless* of whether you own it or are leasing or some other way sharing it.

In short, we call on designers, businesses, and leaders alike to expand their understanding of circularity and embrace this broader, more holistic view of what value is and how it's created. As Ramírez and Mannervik show, organizations can choose to go beyond simply embracing circularity and instead build new value-creating *systems*. This necessarily represents a more fundamental, more holistic way of thinking about value, but ultimately an even more relevant and expansive one. Just as we do not have to accept the dominance of a single model of technological innovation—that of Silicon Valley—we do not have to continue to adopt a narrow model of value creation.

And let's not forget, the stakes have never been greater. "The need for true innovation has never been more profound than now," says Michael Braungart, the cofounder of the influential "cradle-to-cradle" movement. "We are capable of creating high-quality circular alternatives which are beneficial for humans and nature. If the future can be positive, why choose differently?"

This expansion requires us to make arguably the biggest leap: to stop taking the individual human as our point of departure in all things design, but instead to think about our human habitat. In 1965, Adlai Stevenson, the US ambassador to the United Nations, visited Geneva to speak about international development. His message remains as relevant as ever: "We travel together, passengers on a little spaceship, dependent on its vulnerable reserves of air and soil; all committed, for our safety, to its security and peace; preserved from annihilation only by the care, the work and the love we give our fragile craft."

This expansion exposes the scale of the problem and the fundamental nature of all the solutions that follow: we are the crew of Spaceship Earth, and if we are to correct our course, we need to start working more harmoniously together, for the greater good.

EXPANDING HOW VALUE IS PROTECTED

How we create value is one thing; how we protect it is quite another. Historically, intellectual property (IP) has been the juridical framework that captures "added value" and ensures that those individuals or organizations that create valuable new ideas, concepts, brands, and inventions also benefit financially. That asset is IP.

Alphabet (Google's parent company), Amazon, Facebook, and Apple have all cashed in on that inherent design value. IP protects the foundational value of the four tech giants, as well as of many of the world's most significant businesses. The value that design and similar creative disciplines bring to the creation of IP can hardly be exaggerated. We acknowledge that businesses need to secure that value in one way or another.

But in recent years, the paradigm of IP rights has been challenged in several ways. The finer aspects of criticism of IP are beyond the scope of this book, but in brief, the critique has circled around two arguments. First, the term and logic itself: the notion that one can own an idea or concept in the same way as one can own a physical property or piece of land is, to some, problematic.

EXPAND

In the physical world, ownership rights imply that once something is owned, it cannot be used by others without infringing and lessening its value to the owner. But an idea shared with another human (or organization) is not diminished or altered; in fact, the idea may be *even more valuable by being shared*. A second critique is that IP *locks in* value, for instance in big tech or big pharma, and kills competition and innovation. By holding on to ideas for too long, the premium placed on the rights of the inventor or originator is out of balance with the value to the economy and society of sharing the idea openly.

For these reasons alone, we believe it is worth expanding how we think about protecting value. Not because we should not necessarily embrace traditional approaches to the protection of ideas, but because there can be other relevant approaches that enable innovation to be valued and innovators to be rewarded, and that benefits society as well.

In the following section, we'll entertain two directions that can expand intellectual property as a means of protecting and securing value-creation. One is the open-source movement, especially Creative Commons. The other is the question of alternative, culturally based approaches that embrace community and geography.

The idea of creating value by opening up knowledge is not new. Some of the founders of the early internet were proponents of a radical new culture; as Stewart Brand purportedly told a 1984 hackers' conference: "Information wants to be free." This statement later became a more direct critique of the patent system as one form of protecting knowledge. It also foreshadowed the movement, first in software development, later more broadly in business innovation, around open-source and open-innovation and open-business models. The idea was that

EXPANSION 4—VALUE

instead of actively seeking to protect the use of bespoke knowledge, more value could be created (for the individual or business originating the content too) by sharing it, more or less freely. Today, open-source software constitutes a major fraction of the business models of established players like Microsoft, and is foundational to successful enterprises such as Linux, a widely used open-source operating system, and Red Hat, a provider of open-source technologies.

The benefits of opening up, rather than protecting ideas, is twofold. First, many more actors—even entire communities or ecosystems—can access the idea and improve it. Through this type of connection, solutions simply get qualified faster, and new, even better, ideas emerge. The other benefit is speed. Rather than waiting for legal processes to lock in an idea or concept, it can be brought to the market immediately. In some industries (like software development), this can be a huge advantage. In others (like pharmaceutical research), it may not be, since firms in this sector need to recapture years of major R&D spending.

One way to accommodate more flexible models between "open/free" and "closed/protected," is the Creative Commons framework. Run by a nonprofit organization, Creative Commons provides standardized models for individuals and organizations to allow others to use their work and content. Conditions vary, from the CC0 (CC Zero) dedication tool that essentially sets content entirely free in the public domain to be changed, mixed, and used commercially, to CC BY-NC-ND, a relatively restrictive format that only allows credited, noncommercial uses of the original content in unaltered form.

In short, both open source and Creative Commons represent one expansion of the way that knowledge is handled to ensure value creation.

EXPAND

Another potential expansion takes us into the domain of culture and geography. Our argument here is that to truly move from a monotechnological mindset to a world with rich technological diversity, IP cannot be culturally biased. Take, for example, African design, which is largely grounded in the continent's heritage and exposed to contemporary international aesthetics, satisfying local demand, attracting international attention, and shaping new African identities. Outside of exporting contemporary African iconography, there is another type of successful African design approach on the continent—one that fuses indigenous knowledge and modern technology. This approach often makes the resulting designs cheaper, more sustainable, and easier to adapt locally. The work of architect Francis Kéré is a good example of such an approach—in his words, "innovating with mud" based on traditional techniques from Burkina Faso. He has reappropriated traditional building techniques to further develop the materials and building techniques for contemporary use.

One challenge facing this indigenous design approach is how to protect the IP. On the African continent, IP is traditionally considered as belonging to the community and not private to individuals, groups, or companies. But today's IP laws are ill-suited for protecting such knowledge because they are usually based on Western norms such as individual rights and commodification, strictly considering the commercial use of knowledge.

Indigenous knowledge, meanwhile, has been built and passed on from one generation to the next by word of mouth, for the betterment of their societies. Folkloric knowledge and collective inventions are often passed on by elders. The bearers of traditions, elders, let alone the dead, have little significance in IP law as seen by Western scholars. In the West, IP must be new, individual, and innovative to qualify for protection. These requirements make it difficult for indigenous designs developed with a sense of timeless community to qualify for protection.

We must therefore expand what qualifies as intellectual property to protect societies and nations, such as those in Africa, from infringement and let them grow new value. The principle of the appellation system for wine *terroirs* might be the closest Western concept that could be adapted. A terroir's crops have qualities. These environmental qualities are said to have a character that can be protected by an appellation system. Appellation systems, such as the French AOC system, for instance, have developed around the concept of "unique wines from unique areas." Perhaps we could explore how the concept of appellations for *terroirs* would best translate into patents for indigenous design practices. Judging the value of design in today's business environment, you can't just think of Silicon Valley, Scandinavia, northern Italy, and other mature design regions. When what legally regulates value can be expanded, so can the value creation for sure.

On a side note, finding new ways to protect IP could also potentially enable social media companies to make their algorithms more transparent and accessible for regulation or supervision without losing their competitive edge.

Thinking about value also requires us to think multidimensionally—to look through and beyond what a product is or can do, now and again, today and tomorrow, and so on. Arthur Huang, the founder of Miniwiz, cites a cardboard hot dog holder as an example of how to expand our thinking in this way. "We use it for twenty minutes to keep ketchup off our fingers," he explains. "But the material is agnostic. It doesn't care what it was in a previous life. So, can we turn that twenty-minute use into a multiyear second life? Into furniture? Can we turn that seven-year furniture into a seventy-year construction material, like a brick or tile? Can we extend the life of a material through different life cycles, so it continues to be productive and valuable within a new life cycle, and to become even more valuable in the next life cycle?"

While ostensibly about value, this kind of thinking is a bridge to the topic of the next chapter—which makes the case for expanding our thinking about *dimensions*.

6

EXPANSION 5— DIMENSIONS

The mind, once stretched by a new idea, never returns to its original dimensions.
Oliver Wendell Holmes Sr., *The Autocrat of the Breakfast-Table*

All fiction that does not violate the laws of physics is fact.
David Deutsch, *The Beginning of Infinity*

Garry Kasparov's 1997 loss to IBM's supercomputer Deep Blue wasn't just the first time a computer beat a reigning world chess champion under tournament conditions. It was also a milestone in AI's pursuit of human intelligence. Another watershed moment came in 2016, when a program developed by Google's AI division DeepMind defeated the eigtheen-time world champion Lee Se-dol in a best-of-five game of Go, the ancient Chinese board game that had long been considered impossible for computers to play at a world-class level.

EXPAND

Se-dol's defeat was a harbinger of sorts. Artificially intelligent technologies are developing exponentially and enabling algorithms that can solve increasingly complex tasks. Indeed, robots are rapidly mastering everything from cooking and medical diagnosis to financial and legal services—leading to inevitable concerns about mass unemployment, as robots take over more and more jobs, as well as to fears that we may lose our sense of purpose.

In a way, though, what Steve Cramton and Zackary Stephen did a decade earlier is much more significant. Having met at a chess club in New Hampshire, the two men spent several years honing their skills. Stephen, meanwhile, was especially keen on chess programming—and in 2005, he and Cramton entered a freestyle tournament that attracted several teams of grandmasters aided by computers.

The tournament was played online through the servers of Playchess.com. Cramton and Stephen had three computers crunching the numbers, as well as a database they had developed that showed which of the two of them typically did better when faced with similar situations. They also knew how to play this relatively new form of chess. "We had really good methodology for when to use the computer and when to use our human judgment, that elevated our advantage," Stephen told the BBC a decade later.

Relatively unknown in the world of competitive chess, Cramton and Stephen won the tournament, defeating grandmasters and computer programs along the way. Their victory proved that certain human skills remained unmatched by machines when it came to chess—and that using those skills cleverly and cooperatively could make a team unbeatable. Indeed, the larger point about Deep Blue's defeat of Kasparov is not that a human can be beaten. It's that a human playing *alone* can be beaten. As Cramton

EXPANSION 5—DIMENSIONS

and Stephen later showed, a computer is no match for a human playing *alongside* a computer.

In a sense, this story epitomizes the expansion that we explore in this chapter. It concerns the dimensions of design. Cramton and Stephen's feat was to straddle the human and digital dimensions, make the most of both, and thus create a winning approach.

But we need to expand our thinking about dimensions in even more ways.

In this chapter, we explore how the parameters of design have expanded from what we might describe as "medium" in scale (physical artifacts, graphic communications, and so on) to concepts that are much bigger in scale such as policy, organizations, services, and systems—even Earth's climate systems and, potentially one day, those of our nearest neighbors in the solar system.

At the same time that design is expanding into the realm of the ultra-large, the macro—encompassing everything from climate engineering to inhabiting Mars—so too is it expanding (if that is the right word) into the realm of the ultra-small, the micro, the electrons and the kind of supramolecular manipulation enabled by nanotechnology.

That's not all, of course. In recent years, the parameters of design have expanded from the material world to include the digital one. More and more designers manipulate bits and bytes rather than atoms and molecules. They are designing AI, algorithms, and applications—and building new, digital worlds that coexist and interact with our physical one. (In manufacturing, they design fully connected "digital twins" of products that exactly mirror their physical counterparts, allowing for sophisticated diagnostics, simulations, and optimization of the product.) They're designing for a world in which everything and everyone will soon be connected. Intel, the US microchip maker, estimates that more than two hundred billion devices are now linked via the "internet of things." That equates to twenty-six devices—phones, gadgets, vehicles, home appliances, and so on—for every person on the planet.

Moreover, design straddles the physical and digital worlds in ways that were once the purview of science fiction. Consider, for instance, the role that emerging technologies such as virtual reality (VR), mixed reality (MR), and augmented reality (AR) can play in design, and how designers might design for a world in which technology is becoming increasingly interactive and much closer to our bodies thanks to the growth of natural interfaces such as voice-, touch-, and gesture-enabled technologies.

Last but not least, the realm of design has expanded into the realm of artificial intelligence. Most questions about AI concern its inevitability. Will—or rather *when* will, as the Singularitarians would ask—AI replace human intelligence? But there are other important questions worth asking. For example, what are the consequences of AI increasingly *augmenting* human intelligence, and vice versa? And as we increasingly mesh human and computer intelligence into something more powerful than either alone—and as our physical and digital worlds change how we live, work, and play—what are the opportunities? The combination of human and artificial intelligence in design-dominated fields suggest a fruitful future. A future that allows us to move beyond the technological determinism of the past few decades and imagine many more alternative directions of innovation.

We begin our exploration of this topic, however, with a common household object and an example of how the parameters of design can be expanded from what might be described as "medium" in size and scale.

EXPANSION 5—DIMENSIONS

EXPANDING THE DIMENSIONS OF DESIGN

Once upon a time, if you felt cold, you put on another layer or turned up the heating. Then the thermostat came along and ensured that if the mercury fell, your home or office would be kept at a pleasant ambient temperature. In terms of dimension, the thermostat is a medium-size design solution. Tomorrow's solutions can be found at both ends of that scale. Let's start by considering the small one.

"Wouldn't it be cool if your clothing could adapt itself to the environment you're in?" This is not a line of science fiction or the breakfast-table musings of a child with a mind full of ideas, but most likely a question posed at some point by one of the "mischievous scientists" at Otherlab, the San Francisco–based research-and-development firm that works across a number of fields, including robotics, solar energy, energy storage, and medical devices.

Indeed, some of Otherlab's work typifies how the parameters of design can be shifted in the other direction—not into the realm of the macro, such as a district heating system, but into the domain of the microscopic. Together with Kestrel Materials, Otherlab has developed "textiles that do something delightfully unexpected: they move and change shape, increasing their thickness and insulation in response to the cold." The aim is to make clothing that is so "thermally comfortable" that the people wearing it don't need to touch the thermostat. In other words, the clothing would "broaden the range of comfortable temperatures for building occupants," meaning they could reduce energy use.

The development of this technology has taken more than a decade and is firmly on the bleeding edge of design. "Multiple rounds of additional innovation and insight have been required," wrote Kestrel's CEO Brent Ridley in a Medium post in 2016. "Some of the techniques we are using now we didn't know about 15 years ago. What we are doing now only came about through an extended period of unearthing ill-formed ideas, extracting from them what we could, and moving on."

Otherlab and Kestrel are by no means alone in this field. Nanotechnology is increasingly being used in clothing design to create garments that are waterproof, microbicidal, UV-blocking, or antistatic. For example, the clothing brand Organic Basics uses silver nanoparticles in a range of underwear because of their ability to kill bacteria and fungi and prevent nasty smells.

The point here isn't that an emerging form of technology is being applied to fashion—or, for that matter, to food, to medical devices, to construction, or to any number of other industries. It's that within the physical realm, the dimensions of design continue to expand and now encompass the level of the supramolecular, the ultra-small. As we'll see now, those dimensions are expanding at the other end of the scale too.

Danfoss has come far since Mads Clausen launched the company in his parents' farmhouse in Denmark in September 1933. It rose to prominence making thermostatic expansion valves for refrigeration systems as well as the world's first radiator thermostats for regulating heating. In other words, it made physical products, some small enough to hold in your hand. Today, the company is a leading manufacturer in the climate-technology space, making products and services that are used in cooling food, air conditioning, and heating buildings, among other things.

The eighty-five-year-old company is also leading the charge in the field of district heating, which involves centrally produced energy being distributed through pipes to provide hot water and space heating in residential and commercial buildings. In the residential building, Danfoss's digital products can learn when the individual dweller will need heating and cooling. That's a huge span and shift in dimension, and a reflection of its ambition to tackle the UN's Sustainable Development Goals. Indeed, what Danfoss has achieved in the field of district heating typifies the expansion of design in scale or parameters from the small or medium-sized to the very large.

Danfoss is driven by the belief that "district energy is the most sustainable approach for mitigation of climate changes and improving energy efficiency"—in particular, that district heating has a crucial role to play in lowering carbon-dioxide emissions and optimizing energy consumption in cities. For example, heat that otherwise has little or no value in one place—such as industrial surplus heat—can be transformed into a high-value commodity in areas with high heat demand such as homes and businesses.

Several trends are driving the development of district heating, including a shift from single-source to multisource energy and from fossil fuels to renewables and surplus energy. The main challenge is maximizing network and production efficiency while improving the quality of supply. As part of its multifaceted approach to improving district energy, Danfoss has redesigned its supply network. In particular, it has addressed factors that were keeping the cost of production high, such as water and heat losses. Its solution? To verify capacity and assess impact on the whole network, and to avoid overdesigning.

Danfoss's work in district heating is the bedrock of EnergyLab Nordhavn, a visionary new energy system in northern Copenhagen. The winner of a Danish Design Award in 2019, EnergyLab Nordhavn is a "living laboratory" to show how electricity, heating, energy-efficient solutions, and transport can be incorporated into an intelligent, flexible, optimized system.

According to the jury, the project won because "design is also about collaboration and about creating synergies to achieve a common goal. Even if the end result is invisible, a design-led approach is invaluable. Energy is so important to the future of the planet that efforts to work collaboratively across energy sectors are a radical approach."

Here, then, are two ways to think about how to expand the dimensions of design. While Otherlab's work potentially eliminates the need for certain thermostats because its clothes will respond to our surroundings and keep us warm or cool, what Danfoss has achieved with district heating shows how design can be applied to something as vast as a regional energy system. They are also examples of technological diversity at play and represent fundamentally different visions of the context we live in. Do we design thermostats for individuals responsible for their own well-being, or do we design them for groups, cities, or societies? Are these technologies expressions of free, individual citizens guided by Adam Smith's invisible hand, or parts of society's cooperative plan for a better world?

The next question is: How big can a system get? Or what are the biggest systems we might be able to design? Perhaps the most common form of large-scale design project today is the city. Urban design has become its own distinct field, with legions of city planners making the pilgrimage to cities such as Copenhagen to learn how to better design their streets and parks and cycle lanes. But why stop there? In the time chapter, we discussed how some governments have long-term plans to redesign their entire societies. In fact, how about entire continents? Take heating as an example. What Danfoss has done in the field of district heating provides an example of how we might think about and design for the city level. But what if we took *that* to the next level too?

EXPANSION 5—DIMENSIONS

When we think about future energy systems, electricity usually comes to mind. We picture smart power grids that integrate and deliver electricity for different purposes, from powering household appliances to charging electric vehicles. Yet smart power grids cannot dramatically reduce the amount of energy wasted at the source. In the World Economic Forum's Future of Europe working group, Jens Martin proposed creating the power grid's twin—a smart thermal grid.

Supermarkets today produce huge amounts of excess heat to keep their refrigeration systems running 24-7. Similarly, many factories and industrial buildings need permanent cooling or heating systems. And any excess heat is simply released into the sea and air. What if we could retain that excess heat and direct it to where it is needed? What if buildings, supermarkets, and factories could exchange surplus heat directly with each other or distribute it to residential buildings? All we need is the right system. Most countries have all the energy they need for heating and cooling; they can't redistribute it.

A smart thermal water grid would enable all that, and more. By expanding and integrating existing district heating infrastructure, it would allow us to use the excess heat from any wasteful source—power plants, chemical plants, industrial refrigeration systems, and the like—at scale. Moreover, studies show that once such a water distribution system is up and running, it's cheap and effective to add renewable energy sources such as solar, geothermal, and energy from biofuel plants. A smart thermal water grid would also allow energy providers to anticipate and respond to surges in demand (as when a majority of people living in a city switch on their hot shower at about seven o'clock in the morning).

More efficiency would mean that less fuel would be needed to generate energy. That, in turn, would mean lower costs for the provider and cheaper prices for the customers. The new energy infrastructure system would create new jobs in design, construction, and management. It would also pave the way to greater energy independence—and energy security if we could get all the heat we needed from our own surplus and integrate limitless renewable sources into our smart grids.

As it happens, Danfoss is part of a team developing aspects of this idea in Copenhagen and Sønderborg in Denmark, in Anshan in China, and in other cities around the world. And it is no coincidence that Europe is taking the lead on energy efficiency because the heat wasted across the continent is more than that required to heat all of its buildings. A pan-European smart water grid would have a huge impact on the environment and on the economy of the world's biggest marketplace. It also typifies how we can easily start thinking about expanding the dimensions of design.

EXPANDING INTO GEO-ENGINEERING

Continuing the theme, why stop there? Having unintentionally altered Earth's climate for a couple of centuries, some say we need to deliberately "hack the planet" in a bid to limit temperature rises to 1.5 degrees Celsius above preindustrial levels. It's called geoengineering—large-scale intervention in Earth's climate system—and some climate experts now believe we may have to use it as a temporary remedial measure as we head toward dangerous levels of global warming.

For instance, the authors of the IPCC's alarming study on global warming said there was strong agreement that the "injection of millions of tonnes of sulphur dioxide into the stratosphere" could help limit temperature increases to 1.5 degrees

Celsius above preindustrial levels. Other strategies for lowering Earth's temperature include blocking sunlight, planting a billion trees, and enhancing the planet's ability to absorb carbon dioxide.

As author Eli Kintisch told *Wired* in 2010, "Scientists have proposed ways of intercepting solar radiation at every single point from the surface of the earth by whitening roofs or brightening the ocean's surface itself with tiny bubbles, to brightening low-lying and high clouds, to one of the most radical and discussed geo-engineering techniques: adding particles called aerosols to the stratosphere."

China, meanwhile, has become one of the world's leading users of rain-making technology. According to the *Guardian*, in a bid to induce rainfall in drought-plagued regions, it uses a technique that "involves dropping dry ice or silver iodide on cumulus clouds to induce or accelerate precipitation above reservoirs and rivers." And in 2009, it employed "cloud-seeding aircraft to intercept rainclouds" that threatened to ruin the communist party's sixtieth celebrations in Beijing. Not much good for the environment, though.

The dimensions of design, however, are no longer even limited to the Earth. Even as we wreck this planet, we have started to turn our attention to our nearest neighbor. Architect Bjarke Ingels is behind plans for a "space simulation campus" near Dubai where scientists will work on "humanity's march into space." According to Dezeen, the four geodesic domes comprising the experimental Mars Science City will recreate the conditions of the Red Planet, and a team of scientists will work for a year in "laboratories dedicated to investigating self-sufficiency in energy, food and water for life on Mars."

The Mars Science City is just one of the many eye-catching designs proposed as part of the race for space colonization.

According to Dezeen, others include "a space suit designed to be worn by travelers embarking on the 80- to 150-day trip to Mars that Elon Musk plans to launch in 2022 through space exploration company SpaceX" and a range of designs for housing on the Red Planet, including "a submission by Foster + Partners that proposed the use of semi-autonomous robots for construction."

Of course, plans for the long-term colonization of Mars rest on the assumption that we will be able to "terraform" it—that is, create an Earth-like or habitable environment there. Last year, however, NASA poured cold water on plans to inhabit Mars. A NASA-sponsored study indicated that terraforming the planet could not be done with currently available technology. "Any such efforts have to be very far into the future," the space agency concluded. In the meantime, of course, we can still have grand designs for the Red Planet.

EXPANDING INTO DIGITAL

Scale isn't the only way in which the dimensions of design have expanded. It has also expanded beyond materiality—into the realm of the digital, the world of bits and bytes, not atoms and molecules. And as Cramton and Stephen's victory in the advanced chess tournament heralded, the morphing of the physical and the digital is growing in relevance.

To understand quite how we got to the point of increasing convergence between the physical and digital realms, consider how our interactions with computers have changed. There was a time when we had to know computer code in order to be able to use them. The graphical user interface changed that. For the first time, regular people could use computers at home and in the office. Initially, they were chunky machines that took up lots of space, but in time, they got faster, smaller, and increasingly closer to our bodies—first as laptops, later as tablets, and then as smartphones.

Now we are moving into the realm of wearable technology—technology that's in our ears, over our eyes, under our skin. (Think of the growing movement of people who "hack" their bodies and implant microchips, seemingly allowing them to unlock buildings more easily, or, as in Sweden, pay for a train ticket.) One day soon, it will be common to find technology within our bodies, if not our brains. Integrating product experience to include voices and visual is already an integral and growing part of product design. Think of the rising popularity of Alexa or Google Home. (Apple knows which way the wind is blowing. Five years ago, it hired acclaimed Australian designer Marc Newson as senior vice president of design under Jony Ive, who led industrial design as well as Apple's so-called Human Interface software teams.)

Soon, too, we may see design created with even more of our sensory dimensions integrated into products. Today, perceptions of various designs and products are already a combination of at least two senses but mostly audio, visual, or touch/feel. The next generation of product design will see our sense of smell become integral as well, and we'll see that creating the best integrated sense experiences will open new frontiers. At the current extreme is Neil Harbisson. Legally recognized as a "human cyborg," he has an antenna implanted in his skull that translates electromagnetic radiation, such as color, into his sensory system as physical vibrations.

EXPAND

EXPANDING INTO VR

Another blend of the physical and the digital is virtual reality (VR). The immersive technology has growing applications for designers. For example, it has already been incorporated into the work of the leading Danish architecture firm Henning Larsen. "In the earliest stages, the end user of the future building can already get not only a visual impression of the building but also use their other senses for an accurate preview of the space," Jakob Strømann-Andersen, the firm's chief engineer of sustainability, told Arch Daily. "Architects can then make adjustments accordingly based on the user's evaluation of the room. We can alter acoustic settings, light settings and so on, so the indoor climate of the space will fit the end user. We can ask people how they perceive the space way before we hit the site, or even glimpse the rendering. It can qualify the decisions we make as designers in regard to important human aspects like acoustics."

Then there's augmented reality (AR). Unlike VR—which puts physical entities into a virtual space—AR involves virtual entities being made visible to us in the real world through the lens and display of a smartphone or other computer. Perhaps the best-known example is *Pokémon GO*, which the Japanese gaming giant Nintendo launched in 2016. Simply put, it's a game in which players go out into the real world and try to "catch" colorful avatars known as Pokémon—who they can find using GPS, see through their phone's camera, and catch by taking a photo of them.

Arguably the most successful and best-known melding of the physical and virtual worlds, *Pokémon GO* had grossed over $3 billion in revenue and been downloaded a billion times by 2019. It is credited with promoting physical activity—if you've ever spotted groups of young people running around with outstretched smartphones, chances are they are playing the game—and with boosting the foot traffic of businesses that forked out a tidy sum for Pokémon to be placed conveniently nearby.

For what it's worth, Nintendo epitomizes the expansion from one dimension to another. The company was founded in Kyoto in 1889 as a business producing and marketing playing cards. Despite pivoting several times in the twentieth century—Nintendo variously ran a taxi company, a chain of "love hotels," and a food company—its fortune changed in the 1970s when it began to make arcade games. By the 1990s, it was selling games consoles, such as the Game Boy and the Wii, before making its first foray into AR.

Needless to say, there are increasing numbers of applications using AR. Nintendo's *Pokémon GO* became the benchmark, and competitors such as LEGO and *Minecraft* are investing heavily in AR games. The BBC's *Civilizations AR* app allows you to locate, rotate, and resize historical artifacts; Ink Hunter lets you "apply" a tattoo to your body and see how it looks without actually going under the needle; and IKEA Studio allows you to "place" the Swedish company's furniture in your living room or bedroom and "see" what it would look like there. IKEA Studio is perhaps the best use case to date, and launched in 2018.

The expansion of design into the hybrid realm of the physical-digital also has military applications. For example, the US Air Force has successfully tested a helmet with physiological monitoring capabilities. "Its heads-up display shows different information based on how the pilot is feeling and other factors," the website Defense One reported in 2017. "The goal is to give every pilot a slightly different experience based on their unique physical and mental strengths and weaknesses, as well as their physical condition at the moment. Lab researchers and contractors anticipate it will guide the design of the next U.S. fighter jet, to be launched between 2025 and 2030."

Unlike VR, which involves physical entities entering virtual worlds, AR involves placing virtual objects in the real world. It also requires us to look *through* the screen, rather than at it. Instead of interrupting our experience of something real, AR engages with and augments it. And it's only getting bigger. "AR is going to take a while, because there are some really hard technology challenges there," said Apple's CEO Tim Cook in 2016. "But it will happen, it will happen in a big way, and we will wonder, when it does, how we ever lived without it. Like we wonder how we lived without our phone today."

EXPANDING INTO SOCIETAL CHANGE

There are other ways in which the physical and digital worlds are merging, in which the latter affects the former, where the digital dimension is invading our lives not in the context of a cute Japanese game but in a more serious, societal one, and where the data created by our real-world actions can in a sense be turned against us and used to influence our future actions, not only for commercial purposes but for societal and political ones.

Consider the social credit system being rolled out by the government of China. In theory, when the rollout is complete, every citizen and business in China will have a social credit rating reflecting their economic and social reputation. The intention of the system, it appears, is to increase trust and sincerity in Chinese society and to incentivize trustworthy behavior through penalties and rewards. A government document about the system from 2014 says the aim is to "allow the trustworthy to roam everywhere under heaven while making it hard for the discredited to take a single step."

Offenses include failing to pay taxes, spreading false information, taking drugs, using expired tickets, smoking on a train, or walking a dog without a leash. Getting a traffic ticket, for example, would dock you five points. But points can also be gained. Committing a heroic act, say, or helping your family in difficult circumstances would increase your social credit score by thirty points. (It will be interesting to see whether people "game" the system. For instance, when a kindergarten in Israel began issuing fines to parents who picked their kids up late, some parents elected to accept the fine and fetch their kids even later—effectively paying the kindergarten for an hour's babysitting. Might some Chinese citizens try to gain thirty points precisely so they could, say, light up in a smoke-free zone?)

In any case, China's carrot-and-stick system has already meted out draconian measures to some citizens. In March 2019, the *Guardian* reported that China had banned millions of "discredited" citizens from buying plane or train tickets. In fact, according to the National Public Credit Information Centre, by 2019, courts had banned citizens from buying flights 17.5 million times. The center also said that 128 people were prevented from leaving China because of unpaid taxes, adding: "Once discredited, limited everywhere."

The point here is that the digital system that China is designing is aimed at manipulating the behavior of both people and institutions. If it succeeds, it will surely represent the ultimate control by a central power—and one executed within the digital world with consequences in both the physical and digital (among the consequences of ending up on the government's blacklist: a travel ban, a frozen bank account, denied access to luxury hotels, and—a veritable sign of the times—having your social media accounts shut down).

As both machines and the digital realm grow more powerful and more meaningful in our lives, we will need to understand and harness them—and use them positively and creatively. Just because something *can* be designed, we should ask whether it *should* be designed. Does a new solution point to a better world, one that we'd like to live in? As Garry Kasparov said, "We need intelligent machines to help us make our biggest dreams come true. There is only one thing people can do, and that is dreaming—so let's dream big."

But it's the combination of human ability and technological ability that is the most exciting. Potentially, we can look forward to extracting immense value from future human-machine teams. For example, the leading Danish architecture practice 3XN uses AI in architectural design to create the first blueprints for new buildings.

Of course, that invites the question: What happens to the roles of designers or innovators when their creativity and problem-solving abilities can be artificially assisted? Increasingly, we will design and create together with artificial intelligence. Does it make it less human? Less interesting? Less impressive? Less creative? Less iconic?

So far, AI-assisted design has mostly been used for enhancing structural abilities. Human-machine cooperation seems fruitful. The high-end Italian furniture brand Kartell unveiled a part-AI, part-human-designed chair at Salone del Mobile in Milan in 2019. (Superstar designer Philippe Starck provided the flesh-and-blood part.) Dubbed the AI Chair, it is made with 100 percent recycled material and retails for as much as six hundred dollars.

The chair is just one example of how AI-assisted design might transform the look and feel of everyday objects. Already, many simple brackets and components redesigned thanks to AI-assisted design are used in the vehicles, bikes, and planes that propel us.

EXPANDING INTO TOMORROWLAND

In late 2019, Lee Se-dol announced his retirement from professional Go competitions. In an interview with the South Korean news agency *Yonhap*, the eighteen-time world champion said his decision was impelled by the rise of AI. "With the debut of AI in Go games, I've realized that I'm not at the top even if I become the number one through frantic efforts," Lee said. "Even if I become the number one, there is an entity that cannot be defeated."

He was referring, of course, to AlphaGo, the AI system made by the Google-owned artificial intelligence company DeepMind. Three years earlier, Lee had predicted that he would beat AlphaGo in a "landslide," only to lose four-to-one. He subsequently apologized to the South Korean public, saying he had "failed."

Lee, who remains the only human to have beat AlphaGo in a tournament setting, knew which way the wind was blowing and decided to call time on his professional career.

Still, if recent developments in artificially intelligent *chess* are anything to go by, he may yet revisit his decision. In August 2020, a team in North America announced that it had developed an AI chess program that doesn't necessarily seek to beat humans. Rather, it's trained to play *like humans*. The point? A more enjoyable game of chess, for one thing, but also an insight into how computers make decisions differently from humans—and, in turn, how that could help people improve at chess.

"Chess sits alongside virtuosic musical instrument playing and mathematical achievement as something humans study their

EXPAND

whole lives and get really good at. And yet in chess, computers are in every possible sense better than we are at this point," explained Jon Kleinberg, the Tisch University professor of computer science and one of the researchers behind the AI program. "So, chess becomes a place where we can try understanding human skill through the lens of super-intelligent AI."

The purpose of their research is of course less about helping people play better chess (as nice as that is) than it is about exploring how to design AI systems with human collaboration in mind. As Melanie Lefkowitz of Cornell University put it, "In many fields, AI can inform or improve human work—for example, in interpreting the results of medical imaging—but algorithms approach problems very differently from humans, which makes learning from them difficult or, potentially, even dangerous." In other words, human-AI chess teams are just the beginning of our joint enterprise with AI.

And so, as design expands in new dimensions, we might want to ask where it could go next. In other words, how might this expansion keep expanding? In terms of scale, the sky *isn't* the limit. Earlier, we flagged the idea that we might one day terraform or colonize our nearest planetary neighbors. How about beyond that? NASA has already confirmed the existence of more than four thousand exoplanets—worlds orbiting stars like the sun—and thousands of other potential candidates may yet be confirmed.

In terms of dimensions, we may yet witness a wholly different expansion. Indeed, might we embrace a completely different path from the digital one we have pursued for the past half century or so? In a sense, you see, the digital world is merely an expression of a form, with binary electronic pixels standing in for the analogue chemicals of yesteryear, and in that respect, it is also just a random pathway—one that is by no means predetermined. Indeed, one of the arguments of this book is that we should reject the notion of technological determinism.

Indeed, the digital medium is merely arbitrary, and the dominant medium of expression and design could one day be chemically based. We could have been living in a different future, for example, where exploration of the chemistry of photography had led us down another path. For instance, as we've mentioned previously, it is not inconceivable that our computers could one day be run on organic material—aka "wetware"—or that they could have already been based on that technology. Likewise, the realm of bioengineering—which applies engineering principles of design and analysis to biological systems and biomedical technologies such as bacteria engineered to produce chemicals and tissue-engineered organs—continues to grow.

Meanwhile, as virtual shared spaces grow and as the worlds of virtual reality, augmented reality, and the internet converge, we may be forced to contend with (and design for) the extension of that convergence, the so-called "metaverse," that is, a future iteration of the internet, comprising shared, virtual spaces linked within a perceived virtual universe. Just a few years back this seemed like pie in the sky. Today investors discuss which metaverse stocks are the best to buy. When Facebook announced it would change its corporate name to Meta, it was with a clear reference to developing a global metaverse.

The field of AI will also keep expanding. The expansion so far involved a shift from human intelligence to computer intelligence to a combination of human and computer intelligence. What might the implications be of AI that could become self-replicating and self-determining? That will certainly entail a further dimensional expansion—into the realm of what we consider to be living or lifelike.

In any case, we would do well to ask not simply how the dimensions in which we design might expand, but at what cost. Not every expansion in dimension is a wise, responsible, equitable, or sustainable one. Remember that we need not chase the chimera of technological determinacy. We believe these developments will require a discussion within the design field about digital ethics.

Take Google Personalized Search, a feature of Google Search that was introduced in 2004 and revolutionized the internet. All searches on Google Search became associated with a browser cookie record. When a user searches for information online using Google, the results they see are based not only on their relevance, but on the search preferences of the user. This has provided more personalized experiences, but it has also created so-called filter bubbles. What one user sees when they Google, say, "cause of climate change" may be very different from what another user sees, depending on their individual search history. This feature may even have been the cause of—or at least have reinforced—the recent surge of fake news.

So, was the creation of Google's Personalized Search desirable? As we expand creative problem-solving into new dimensions, we need to remember another expansion—that of proximity and, thereby, of ethics.

Now, after all this talk of terraforming and exoplanets, for our final expansion, let's return to more familiar ground: the role of the public, private, and civic sectors.

7

EXPANSION 6— SECTORS

For too long, people have acted as if the private sector were the primary driver of innovation and value creation and therefore were entitled to the resulting profits.
Mariana Mazzucato

A concept is a brick. It can be used to build a courthouse of reason. Or it can be thrown through the window.
Gilles Deleuze, *A Thousand Plateaus*

For almost eight hundred years, the most famous attraction in the Danish city of Sønderborg was its medieval castle, which boasts a large collection of regional furniture, textiles, crafts, and art. But that was before the city invited the celebrated American architect Frank Gehry to develop a master plan for transforming its once-industrial harbor. Since 2008, the waterfront has become a vibrant modern district, with attractive housing, office buildings, parks, a multicultural house with a library and art school, and a large conference hotel. Of course, visitors still flock to Sønderborg to see its magnificent twelfth-century castle—but it's the city's eye-catching new waterfront that gets more Instagram likes these days.

Social media success aside, what's especially interesting about Sønderborg's urban regeneration is how many different partners were involved and the cross-sectoral cooperation that was required to regenerate the waterfront. The origins of the regeneration master plan lie in a public-private partnership (PPP) between a philanthropic organization (the Bitten and Mads Clausen Foundation), an investment firm (Bjarne Rasmussen Holding), and the public sector (Sønderborg Municipality).

In particular, the Bitten and Mads Clausen Foundation teamed up with PFA Pension, a private company, to build Alsik—southern Denmark's largest four-star conference and business hotel, with 190 rooms, three restaurants, and a spa and wellness area inspired by "Nordic living." The hotel was designed by one of Denmark's leading architecture firms, Henning Larsen. To minimize its environmental impact, Alsik turned to local hero Danfoss. (Sønderborg Municipality is the company's global home.) Danfoss provided the hotel with energy-efficient technology, such as intelligent control of all engineering plants, low-energy cooling, the reuse of surplus heat and reduced flow temperatures.

Thanks to its new green-tech, Alsik will purportedly reach 76 percent carbon neutrality—and thus play its part in Project Zero, a plan to make Sønderborg and the surrounding region carbon neutral by 2029 through sustainable growth and the creation of "green jobs." Project Zero was launched in 2007 as a joint venture between the city's companies, citizens, and politicians. Like the waterfront regeneration plan, Project Zero was established as a

EXPANSION 6—SECTORS

PPP: the broad investment circle included Danfoss, Sønderborg Municipality, the Nordea Foundation, and two private energy companies.

Public-private partnerships aren't new, of course, but they do typify a growing trend; namely, a re-evaluation of the potential roles, contributions, and interactions of the traditional domains we've used to categorize societal and economic activity—public, private, and civic. For years, the prevalent notion in capitalist countries has been that the private sector creates economic value, the public sector destroys it—or, at best, redistributes it—and the civic sector "simply" distributes funds to deserving causes.

No longer. The boundaries between the sectors are blurring. The private sector is increasingly concerned with the future of the planet and addressing "wicked problems" and the UN's Sustainable Development Goals. The public sector is demonstrating itself to be an increasingly innovative and entrepreneurial value creator and market shaper. And the civic sector is increasingly thinking more strategically and holistically about the impact of its investments, not just the recipients of its largesse.

As the traditional domains we use to categorize social and economic activity blur, we believe an expansion of the roles and contributions of different sectors, and the interactions between them, is increasingly crucial to designing for sustainable change. This chapter is thus concerned with sectors and, in particular, our understanding of the roles of the public and private sectors in design. In it, we make the case for seeing the public and private sectors as complementary sets of resources, each with their own opportunities and constraints; for letting go of industrial society's narrow notions of sectors and silos; and for thinking much wider in terms of ecosystems and value-creating relations among actors, each with their own constraints, interests, and types of value to be exchanged and cocreated. Make no mistake: an expansion of the roles and interactions between the sectors will be crucial to designing for more equitable, sustainable growth.

EXPANDING INTO GREATER SOCIETAL IMPACT

Let's begin by stating the obvious. The private sector ain't what it used to be. In a 2019 survey of global executives by the consulting firm Deloitte, the respondents revealed that they had a genuine commitment to improving the state of the world. In particular, when asked to rank the most important factors their organizations use to evaluate success, more than a third of executives cited societal impact first—more than financial performance and employee satisfaction *combined*. And in the past year alone, almost 75 percent of executives said their organizations had taken steps to make or change products or services with societal impact in mind.

By the same token, when the Danish Management Society held its annual summit in 2019 at UN City in Copenhagen, the primary focus was on addressing the UN's Sustainable Development Goals. For the many businesses present, the question was no longer "*Should* we have a positive impact?" but "*How* do we have a positive impact?". For many businesses today, doing "good" is not incompatible with turning a healthy profit.

Similarly, there has been an explosion in impact investment. It is nothing new that investors can choose to take environmental, social, and governance factors (ESG) into consideration when making investment decisions. However, impact investments go further. Rather than merely assessing ESG factors (which concern how sustainably firms behave), impact investments focus on generating competitive financial results while *also* achieving positive and measurable societal and environmental impacts. Often, impact investments seek to align outcomes with the UN's SDGs or similar sustainability frameworks.

Long a niche philosophy, impact investment is emerging rapidly, even though it is still far from mainstream. The likes of Bank of America/Merrill Lynch, BlackRock, Bain Capital, and TPG Capital have all launched impact investing units in the past five years. The Global Impact Investing Network (GIIN) estimates the current market size of impact investment at $715 billion. That corresponds to roughly one percent of global assets under management.

That number is likely to grow in light of increasing popular support for the idea of investing in firms that make a positive difference to society and the planet. According to a recent study by Allianz Life, almost 80 percent of respondents said they "love the idea of investing in companies that care about the same issues" as them, while 71 percent said they would stop investing in a company if it behaved unethically. For example, when asked about the importance of a variety of factors in making a decision to invest in a company, 73 percent of respondents cited environmental concerns, such as a company's carbon footprint.

In a way, then, more and more businesses are starting to act like the public sector. Which is to say, increasing numbers of enterprises believe addressing one or more of the UN's Sustainable Development Goals to be their raison d'être. These are companies that want to succeed financially—but to do so *sustainably*. They are concerned with the so-called triple bottom line and are as concerned with their social and environmental impact as they are their financial success. To put it another way, these companies all look beyond the next set of quarterly results or the next set of dividend payouts to short-term investors. Instead, they want a healthy planet on which to be able to operate in a hundred years' time.

Similarly, there has been an explosion in startups and social entrepreneurs seeking first and foremost to "make the world a better place" with their product or service. (The phrase has become such a pitch-session cliché that it has even been lampooned on the popular HBO series *Silicon Valley*.) Joking aside, investment portfolios are bursting with clever ideas from startups addressing problems that were once the domain of the public sector.

Take StepJockey, a British startup that promotes health and combats sedentary behavior. Its big idea? Placing smart signs in the stairwells of office buildings and gamifying the use of stairs via an app that can read the signs and record user data. In short, the app allows employers to introduce stair-climbing challenges that can purportedly increase their employees' use of stairs by up to 800 percent.

The app was developed thanks to a £200,000 contract from the UK's Department of Health. (Under the Small Business Research Initiative, businesses can win up to £1 million to develop a new product or service for the public sector.) StepJockey has since secured three times that amount in private investment. Customers include Deloitte, Disney, and Barclays as well as hospitals and public authorities in the UK and overseas.

StepJockey's chief executive, Paul Nuki, said the government contract allowed the company "to properly and scientifically validate the idea with prototypes . . . It allowed us to build and deploy a live product. After that, the fact that government, and in particular the Department of Health, had been a supporter made a big difference on clients. It gave the company a status and a credibility."

EXPANSION 6—SECTORS

EXPANDING INTO THE ENTREPRENEURIAL STATE

StepJockey is a good example of a startup that the public sector kick-started. Not only does it show how the two sectors don't exist in silos, but it also belies the notion that the private sector alone creates and shapes markets. Indeed, the idea that the private sector exists solely to extract economic value is as fundamentally flawed as the notion that the public sector is there simply to redistribute hard-won wealth.

A charge led by the economist Mariana Mazzucato has forced a rethink of the dynamics of private- and public-sector activity and value creation. In a 2011 paper for the think tank Demos, Mazzucato wrote,

> *Across the globe we are hearing that the state has to be cut back in order to foster a post-crisis recovery, unleashing the power of entrepreneurship and innovation in the private sector. This feeds a perceived contrast that is repeatedly drawn by the media, business and libertarian politicians of a dynamic, innovative, competitive private sector versus a sluggish, bureaucratic, inertial, "meddling" public sector. So much so that it is virtually accepted by the public as a "common sense" truth.*

Mazzucato went on to challenge this "minimalist view" of the state and declares that a "far more proactive role is required." In particular, she argued that

> *the role of the government, in the most successful economies, has gone way beyond creating the right infrastructure and setting the rules. It is a leading agent in achieving the type of innovative*

breakthroughs that allow companies, and economies, to grow, not just by creating the "conditions" that enable innovation. Rather the state can proactively create strategy around a new high growth area before the potential is understood by the business community (from the internet to nanotechnology), funding the most uncertain phase of the research that the private sector is too risk-averse to engage with, seeking and commissioning further developments, and often even overseeing the commercialisation process.

In her 2021 book *The Mission Economy*, Mazzucato proposed "missions" to solve some of our biggest challenges, including climate change and income inequality, and to inspire cross-sector collaboration. The idea is that by setting ambitious yet realistic, time-bound, and concrete goals—where the question "Did we complete it?" can be definitively answered—governments can stimulate new market creation and invite private parties to innovate around societal challenges that need to be addressed. In this respect, the public and private sectors could be seen more as a complementary set of resources.

This shift in thinking is apparent in the European Union's new innovation strategy that embraces Mazzucato's mission-oriented framework. In June 2018, European Commissioner Carlos Moedas announced a €100 billion research and innovation program for the next EU's long-term budget. Known as Horizon Europe, the program "aims to keep the EU at the forefront of global research and innovation."

The EU institutions reached a provisional agreement on Horizon Europe in March 2019. Its predecessor was a program known as Horizon 2020, and it also featured notable cross-sector initiatives such as the European Institute of Innovation and

Technology (EIT), an EU body created by the European Union in 2008 to strengthen Europe's ability to innovate. The EIT drives innovation in Europe by supporting entrepreneurs, innovators, and students. It sits at the heart of the continent's largest innovation community, amid universities, research centers, and businesses.

Daria Tataj, a former member of the founding board of the EIT, calls this approach "Network Intelligence" and told us that "high network IQ is the driver of high-performance ecosystems" and is "used by 700+ companies, universities, and best research labs across Europe." The idea is to form partnerships that bring research, education, innovations, and entrepreneurs into trust-based communites to create startups. "Whether in Europe or China, the Silicon Valley model cannot be replicated," Tataj says.

The EU isn't alone in its efforts to break away from siloed, sector-based approaches. Consider the UN Development Programme's sixty accelerator labs. Serving seventy-eight countries, in collaboration with the governments of Qatar and Germany, the labs represent the UNDP's innovative approach to tackling our biggest challenges. The aim is "to find radically new approaches that fit the complexity of current development challenges" in a world in which the speed, dynamics, and complexity of the social, economic, and environmental problems we face are "fundamentally different from previous eras in history."

The labs' watchwords are "collective intelligence" and "partnership." Cross-sectoral thinking is inherent. "Together with partners, the Labs will analyse challenges within local contexts to identify connections and patterns in search of new avenues of work to act effectively in addressing wicked development challenges," the UNDP explains. "The Labs will identify grassroots solutions together with local actors and validate their potential to accelerate development. Solutions can come in many different forms, from a farmer discovering a new way to prevent floods to a nonprofit that is especially impactful."

In a similar vein, the OECD's Observatory of Public Sector Innovation (OPSI) describes public-sector innovation as a "many-splendoured" thing. Innovation itself is multifaceted, OPSI argues, and includes aspects such as "adaptive innovation" (which requires consideration of how an evolved situation changes how something is done) and "anticipatory innovation" (which takes as its premise that emerging possibilities could change what something could or should be). In this light, OPSI is developing a model "to prompt governments to think about why they are innovating and whether they are using the right mix of approaches to achieve their innovation aims." Again, notice the premise: innovation isn't just something governments could do—it's something it should actively do.

EXPANDING THE (EVEN MORE) ENTREPRENEURIAL STATE

The idea that the public and private sectors have little if anything to do with each other is an illusion, of course. The existence of entire industries whose origins were sparked to life by the state belies the notion that the private sector alone creates wealth, jobs, and value, or that it is the sole source of new market creation. In fact, many household names owe their existence to public largesse—which is to say, the generosity of the taxpayer.

Take the arch-entrepreneur Elon Musk. His multi-billion-dollar empire has benefited to the tune of some $5 billion in government subsidies. According to data compiled by the *Los Angeles Times*, Musk's companies Tesla Motors, SolarCity, and SpaceX have together received an estimated $4.9 billion in state support.

The figure underscores a common theme running through Musk's empire: "a public-private financing model underpinning long-shot start-ups," the newspaper explained. "The figure . . . comprises a variety of government incentives, including grants, tax breaks, factory construction, discounted loans and environmental credits that Tesla can sell. It also includes tax credits and rebates to buyers of solar panels and electric cars."

The *Los Angeles Times* added, "Musk and his companies' investors enjoy most of the financial upside of the government support, while taxpayers shoulder the cost. The payoff for the public would come in the form of major pollution reductions, but only if solar panels and electric cars break through as viable mass-market products. For now, both remain niche products for mostly well-heeled customers."

Then there's the nativity story of the internet. In 1958, a year after the Soviets launched the Sputnik satellite, President Eisenhower created the Advanced Research Projects Agency (later known as DARPA), a unit of the Department of Defense charged with developing new technology. It soon set out to create a system that could not be destroyed by a Soviet nuclear attack, which sparked the idea of connecting computers in a network rather than via a single hub. That led to ARPANET, which led to the internet.

And that's not even the whole story. DARPA and other US government agencies also created many of the tools that comprise today's digital economy. For instance, we can thank US government funding for hypertext linking, GPS, and the Unix operating system. When Sergei Brin and Larry Page developed the algorithm that led to Google, they did so with funding from the National Science Foundation. And as Mariana Mazzucato shows in *The Entrepreneurial State*, every major component in the iPhone was developed by the US government, from touch screens to the voice-activated personal assistant, Siri.

DARPA also fueled innovation in driverless cars, running an annual competition to kickstart R&D in the field. "Anyone could compete, their cars had to trundle across 142 miles of desert to get to the finish line, and the winner would take home $1 million," American tech writer Nigel Cameron has written. "Incredibly given where we are now, every single entry in that first round failed to make it. But DARPA persisted, and in 2005 five vehicles crossed the finishing line, with Stanford University taking home the prize."

Danish state support for wind energy is also a case in point. As part of its bid to increase energy self-sufficiency following the global energy crisis in 1973, Denmark introduced subsidies for the construction and operation of wind turbines and biomass plants in 1981. Because it is flat and close to the sea, Denmark enjoys excellent wind resources and often exports its surplus wind energy. Today it is home to the world's biggest wind-turbine maker, Vestas Wind Systems, and the world's biggest offshore wind-park operator, Ørsted. In 2017, the Danish energy minister announced that, after more than four decades of relying on subsidies, the country's renewable energy industry was poised to be able to prosper on its own—sooner than anyone had expected.

EXPANDING INTO HYBRID SECTORS

These examples support the argument that Mazzucato makes in her original Demos paper. "They force the debate to go beyond the role of the state in stimulating demand, or the role of the state in 'picking winners' in industrial policy, where taxpayers' money is potentially misdirected to badly managed firms in the name of progress, distorting incentives as it goes along," she writes. "Instead it is a case for a targeted, proactive, entrepreneurial state, able to take risks, creating a highly networked system of actors harnessing the best of the private sector for the national good over a medium- to long-term horizon. It is the state as catalyst, and lead investor, sparking the initial reaction in a network that will then cause knowledge to spread. The state as creator of the knowledge economy."

While the public sector is starting to play a more entrepreneurial role, in many cases the private sector is starting to operate in domains once thought of as the sole remit of the state. For example, as it shifts its focus to deeper space missions, such as the journey to Mars, NASA is increasingly letting companies perform certain tasks—like launching cargo ships to the International Space Station (ISS). Since 2010, Elon Musk's once heavily subsidized company SpaceX has been quick to capitalize on this new market. It has had seventy-two successful launches of its Falcon 9 rockets, delivering cargo to the ISS as well as launching communications satellites and Earth observation satellites into orbit.

Meanwhile, the ISS itself needs an overhaul—and privatization could be the solution there too. "NASA is trying to develop economic development in low-Earth orbit," said Bill Hill, NASA deputy associate administrator for exploration systems development in 2016. "Ultimately, our desire is to hand the space station over to either a commercial entity or some other commercial capability so that research can continue in low-Earth orbit."

By 2016, the commercial space industry reportedly accounted for a third of all activities on the ISS. "There are already what

I would call experimental things going on," explained James Muncy, founder of US-based space policy consultancy PoliSpace. "You would never build a permanently manned space station in order to pursue these economic opportunities but given that we have the marginal cost of using the space station, to generate new economic activity makes sense." For example, the 3-D printing company Made in Space has launched a project to make optical fiber on the ISS because manufacturing in micro-gravity increases fiber quality. Private space tourism is another fast-emerging commercial market, of course. During the fall of 2021, billionaires Jeff Bezos, Richard Branson, and Elon Musk all whisked non-professional passengers on flights to space (some only on a suborbital path).

EXPANDING THE POWER OF PHILANTHROPY

It isn't only the public and private sector that's changing. Philanthropy is too. Historically, of course, the civic sector was about giving money to seemingly worthy causes (and hoping that splashing the cash would deliver a positive outcome down the line). That kind of thinking no longer cuts it. Today's philanthropists are more metric-driven, collaborative, and concerned with delivering not just societal change but measurable outcomes. You could say they're thinking more like private-sector investors, or public-sector agencies.

As part of this shift, some civic organizations are being increasingly hard-nosed about the recipients of their largesse. Take the charitable fund of the Rockefeller family. Three years ago, it said it would dump all its investments in fossil-fuel companies. The Rockefeller Family Fund was set up in 1967 by descendants of John D. Rockefeller, who made his fortune from Standard Oil—and in its announcement, the fund actually singled out ExxonMobil, describing the world's largest oil company as "morally reprehensible."

But here's the thing. The fund's divestment of fossil fuels represents a shift in thinking within philanthropic organizations, an expansion in their thinking. Where once they gave money to deserving causes while still investing in industries or companies that still made the planet worse off, now they're asking an important question: What if we didn't just give money to good guys, but we actually invested our funds so that it generated impacts aligned much more directly with our philanthropic cause?

Similarly, in the summer of 2019, the Norwegian parliament gave the go-ahead to the country's sovereign wealth fund to withdraw all its investments in fossil-fuel companies: eight coal companies and an estimated 150 oil producers. The world's largest sovereign wealth fund manages $1 trillion of Norway's assets and has been permitted to dump more than $13 billion of investments. The fund will also be able to invest directly in renewable energy projects. The mandate passed by Norway's parliament allows the fund to invest up to $20 billion, starting with wind and solar projects in developed markets.

There is of course a delicious irony at the heart of this shift in thinking. The Norwegian sovereign wealth fund is built on the country's legacy oil earnings—and according to the *Guardian*, it has long argued that failure to cut ties with high-carbon companies could deplete the state pension fund if its investments decline in value in a lower-carbon world.

A similar example of the kind of influence that a philanthropic vehicle can have is apparent in South Africa, where the Bafokeng Community Model is a paragon of resource-related development.

EXPAND

The model is based in the twelve hundred square kilometers of land that the 150,000 members of the Bafokeng tribe inhabit. It is the world's richest source of platinum, and over the years, income from mines on their land have helped the Bafokeng build an asset portfolio worth some $2.5 billion. That, in turn, has allowed them to fund health care and education as well as housing and infrastructure projects. And as part of their community model, Bafokeng villagers get a say in how their windfall is used, with many of their ideas woven into the community's long-term plans to be self-sufficient.

We should, however, pause for a second and consider why this is happening. Why are the sectors starting to overlap more? Why is the state becoming more entrepreneurial? Why is the private sector applying itself more to the world's wicked problems? Why are philanthropists trying to address solutions with targeted investments and collaborative thinking? In short, we believe the answer to all these questions is that growing political and economic volatility, together with technological changes, has fundamentally and forever changed the context within which traditional sectors operate.

In this light, we see two possible scenarios for innovators regardless of their sector. In the first scenario, we are able to predict trends far into the future (think: climate change). In this instance, we largely know what is likely to happen (even if we don't like it), and on the back of our prediction (think: rising temperatures, melting ice, climate migration, etc.), we can start to innovate toward solutions and find ways to address the problems. We can establish tangible missions and start solving whatever needs to be solved to reach them, or at the very least to start on the right trajectory. Our own predictions thus require us to be creative—across sectors.

In the second scenario, we're unable to predict the future, but instead have to innovate in order to discover what we need to respond to (and innovate toward). Here, the challenge is inquiry: to explore and discover problems and frame them in ways that sooner or later allow us to start working on them too.

Either way, we're forced to work across sectors. We can no longer govern or innovate in silos. Circumstances require cross-border, cross-sector, cross-disciplinary creativity, no matter whether we're running a business, a government agency, a research institute, or a foundation.

One inevitable consequence of the blurring boundaries between traditionally siloed sectors are hybrid organizations or platforms. Consider the number of family foundations that run companies in Scandinavia—including global leaders such as Carlsberg, LEGO, and Danfoss. In chapter 1, we noted the dominance of family-owned or foundation-run companies in Denmark operated as a kind of backstop to short-term thinking.

Again, according to a 2010 study, family-run businesses in Denmark see higher profit margins than non-family-run businesses in every sector but one (finance and business). These companies represent the entrepreneurial and profitable management of "private" enterprise by the philanthropic sector. It should go without saying that these foundations are also involved in impact investment in other initiatives and are taking a hard-nosed, data-driven approach to financially supporting everything from scientific research to innovation in children's education. For these foundations, in other words, doing good is not incompatible with doing good business.

Another example of hybridization can be found in the African Leadership University (ALU), "a Mauritius-based institution with an ambitious goal to train the continent's future leaders."

Fleet-footed disruption is the name of the game. In an email to staff in 2018, Fred Swaniker, a Ghanaian entrepreneur who cofounded ALU with Acha Leke, said: "We are a disruptive leadership institution, not an academic institution." Swaniker told the *Times Higher Education* website that he wants to provide "large-scale, unconventional education" to Africa's youth and had concluded that "drifting back to the traditional university model . . . would not help us achieve our vision within the very specific context of Africa," which does not have the "luxury of time and resources" to go down the more traditional route.

To unlock this unconventionality, ALU has turned to investors. In early 2019, ALU announced that it had raised $30 million in a Series B round led by the Bestseller Foundation of the privately held family-owned clothing company Bestseller. "The new funds will be focused on rolling out the organization's lifelong learning centres program, ALX, which opened in Nairobi last year and is set to open in Johannesburg, Lagos, Cape Town, and Casablanca," explained *Quartz*.

Unlike traditional graduate programs, the centers operate out of low-cost setups such as state-funded coworking spaces with career placements and incubator programs for entrepreneurs. Some of the innovations include funding tuition fees by giving royalties of future income to ALU, having students and professionals bounce ideas and techniques off each other, and lengthening semesters to accelerate programs. "In our approach to education, we are focused on ensuring our students don't just learn theory, but also graduate with the set of skills needed to be effective in the real world," Swaniker told us. "This includes offering our students a four-month internship every year, where they acquire practical work experience with major corporations."

EXPANSION 6—SECTORS

Arguably, however, the best-known example of hybridization or a cross-sector platform is the World Economic Forum. It typifies the expansion in thinking described in this chapter—but was, of course, way ahead of its time. Dubbing itself "the international organization for public-private cooperation," the WEF was established in 1971, in Switzerland, as a not-for-profit entity to "demonstrate entrepreneurship in the global public interest."

To do so, it "carefully blends and balances the best of many kinds of organizations, from both the public and private sectors, international organizations and academic institutions." The WEF was arguably the first major organization to recognize the need to facilitate new conversations between business and government leaders, and in time to scale those conversations to include civic society too.

Klaus Schwab, the founder of the World Economic Forum, told us the organization he launched almost half a century ago matters more than ever in the Fourth Industrial Revolution. "Progress is driven by business and entrepreneurial startups, while governments have difficulties coping with the speed," he says. "Traditional methods of creating regulatory frameworks do not work anymore. What's needed is agile governance, which means a continuous interaction between the public and private sector to formulate and relentlessly update human-oriented and society-oriented principles and practices."

To see the kind of impact the WEF can have, also at a local level, look at what happened in the Danish municipality of Faaborg-Midtfyn. In June 2018, it was considering how to help residents get themselves around the municipality more easily. The aim was to increase quality of life in Faaborg-Midtfyn and attract more young people to settle there. Enter the WEF, and in particular the Global Future Council on the future of cities and urbanization.

The group met that summer in Copenhagen and invited executives from international organizations such as MIT, Cisco, and Microsoft to take part. As part of their workshops, the group turned their attention to Faaborg-Midtfyn and ultimately provided the municipality with a slew of innovative ideas aimed at improving the mobility of residents and preparing for the Fourth Industrial Revolution. Today, the municipality continues to work on the project, and its relationship with this WEF expert group helps to give the project clout and attract international companies as potential partners.

EXPANDING—OR IMPLODING—THE SILO

The billion-dollar question is where this expansion goes from here. Arguably, of course, it is less of an explosion, more of an implosion, as traditionally siloed sectors collapse in on themselves or expand into one other.

Some have made the case for entirely new models of organization. Jeremy Rifkin, in *The Zero Marginal Cost Society*, argues that a new economic paradigm known as the Collaborative Commons is fast emerging in the wake of capitalism and will soon transform our way of life. For Rifkin, the Collaborative Commons is where the "oldest form of institutionalized, self-managed activity in the world"—one that "predates both the capitalist market and representative government"—meets the so-called internet of things, or the connection of "everything with everyone in an integrated global network."

"The Collaborative Commons is already profoundly impacting economic life," Rifkin writes. "Markets are beginning to give way to networks, ownership is becoming less important than access, and the traditional dream of rags to riches is being supplanted by a new dream of a sustainable quality of life."

"We are so used to thinking of the capitalist market and government as the only two means of organising economic life that we overlook the other organizing model in our midst that we depend on daily to deliver a range of goods and services that neither market nor government provides," Rifkin writes. He further argues that while the capitalist market is unlikely to disappear entirely, it will no longer dominate or define civilization's economic development. "There will still be goods and services whose marginal costs are high enough to warrant their exchange in markets and sufficient profit to ensure a return on investment," he concludes. "But in a world in which more things are potentially nearly free and shareable, social capital is going to play a far more significant role than financial capital, and economic life is increasingly going to take place on a Collaborative Commons."

According to the WEF Global Future Council on Agile Governance, meanwhile, "there is an urgent need for a more agile approach to governing emerging technologies and the business models and social interaction structures they enable." The solution—agile governance—is defined by the group as "adaptive, human-centred, inclusive and sustainable policymaking, which acknowledges that policy development is no longer limited to governments but rather is an increasingly multi-stakeholder effort."

Others see the appeal in hybrid business models such as the long-established cooperatives of Arla Foods and Mondragon, the agricultural cooperative of over ten thousand milk producers in Denmark, Sweden, the UK, Germany, Belgium, Luxembourg, and the Netherlands. They have the opportunity to influence decisions, and the money that Arla Foods makes is equally split between them. Every year the farmers invest some of their earnings back into the business so it can develop.

Mondragon is a corporate federation of workers based in the Basque region of Spain. The country's fourth-largest industrial group, it is involved in finance, industry, and retail. Its business culture is derived from ten basic cooperative principles, including democratic organization, participatory management, and payment solidarity. Its four corporate values are cooperation, participation, social responsibility, and innovation.

Finally, in *Complexity and the Art of Public Policy*, economists David Colander and Roland Kupers argue against the narratives of market fundamentalism and government control and instead "develop innovative bottom-up solutions that, through new institutional structures such as for-benefit corporations, channel individuals' social instincts into solving societal problems, making profits a tool for change rather than a goal." Moreover, they argue that the central role of the government in this respect is to nurture an ecosystem within which different forms of social entrepreneurship can flourish.

Entrepreneurship—social entrepreneurship in particular—is the connective tissue shared by the public, private, and civic sectors as they adapt to our fast-changing, volatile world. Just as we are finally starting to recognize that the state can be entrepreneurial, so too are many private enterprises increasingly rethinking their guiding philosophy and applying their entrepreneurial nous for the public good as well as for profit. And many philanthropists increasingly believe it makes more sense to spur innovation and entrepreneurship through impact investment rather than by giving away money to worthy causes.

Why has entrepreneurship come to dominate or influence these different sectors? We believe it's because there is a growing need for agency, a growing need for these different sectors to change how they behave. The public sector, for example, is becoming more entrepreneurial in the face of several recent shocks to the system—from the global financial crisis to the rise of populism to the increase in migration, and, not least, the COVID-19 pandemic. Some even speak of a "covid dividend" of enhanced innovation in public institutions in the wake of the global health crisis. Amid the need for action and agency, entrepreneurship is stimulated. And what of the role of design in this? In a way, perhaps, you could say that both entrepreneurship and design are ultimately about creation. That is, they are both about making something that would not otherwise exist were it not for human creativity.

The siloed sectors we've traditionally known are morphing and changing in part because they want to, but also because they have to. They are being forced to change by the volatile world around them. Yes, the public sector is becoming more dynamic, more agile, more adaptive. Yes, the European Union will pursue ambitious missions—and stimulate public and private innovation to achieve those missions. The territories of sectors are expanding—and necessarily converging and absorbing one another. Yes, the world is becoming bolder, more proactive, and more innovative. Why? Because it has to. And, as no problem can be addressed in isolation, the challenge for these sectors will be to find ever better ways to design tomorrow's world together.

Thankfully, there's no shortage of initiatives. Consider, for instance, Impact the Future—"an annual conference that brings together public and private players from more than 50 cities around the world, who are working on solving both local and global challenges through entrepreneurship, innovation and technology." Centered around the seventeen UN Sustainable

EXPAND

Development Goals—with "A Green Inclusive Restart" as its inaugural theme—Impact the Future has what it calls a "bold ten-year mission" and plans "to highlight, celebrate and connect the most promising startups, innovators, corporates and cities that have taken the SDGs into their core business, and help them lead the way for the next ones to come." If that doesn't sound like an expansion of sectors, what is? And the venue for the inaugural conference? Where else but the Danish city of Sønderborg.

8
APPLYING THE EXPANSIONS

> It's a new challenge to see how people can change your look.
> I like words like *transformation, reinvention,* and *chameleon.*
> Because one word I don't like is *predictable.*
> **Naomi Campbell**

> Far out in the uncharted backwaters of the unfashionable end of the western spiral arm of the Galaxy lies a small unregarded yellow sun. Orbiting this at a distance of roughly ninety-two million miles is an utterly insignificant little blue green planet whose ape-descended life forms are so amazingly primitive that they still think digital watches are a pretty neat idea.
> **Douglas Adams,** *The Hitchhiker's Guide to the Galaxy*

A different class of thinking will be indispensable if we are going to change organizations, businesses, societies, ecosystems, and the fate of our little blue-green planet. But there isn't going to be a one-size-fits-all push-button solution. And we can only design a better future and develop tomorrow's truly innovative ideas if we move beyond Bay Area determinism, monotechnological fetishism, traditional management thinking, shortsighted governance, and "human-centered" design. We need to break out of what the artist Olafur Eliasson calls the "dogmatized, manifestoed, siloed, monolith-ified, Disneyfied, McKinseyfied" mindset.

After all, all methodologies, even the most obvious ones, have limits. Too often we think that technology, like science, develops along a given ideal path. But as Paul K. Feyerabend has argued, there isn't a single scientific rule that hasn't been violated by someone. Violations are necessary for progress. Assuming that the scientific method isn't distinct from other intellectual activities, it surely follows that if scientific knowledge emerged thanks to individuals who flouted methodological rules, so must technology. Why, then, would technology be bound by certain rules or follow a given path? And, in a broader sense, beyond technology, why would the solutions we today find obvious necessarily be relevant to tomorrow's problems?

Let's take this thought a little further. We said earlier that design can be viewed as increasing the range of *projected alternatives or possible futures*. For instance, imagine if our familiar alphabetic system had never come to dominate—and that the *quipu*, the method used by ancient Andean civilizations to store and communicate information, had risen to prominence instead. The string-, knot-, and color-based device the Andeans used to record dates, statistics, accounts, and narrative records would now be ubiquitous—at least until some sort of handheld computer became the universal medium. What would that computer look like? How would it function? How would it be designed? Would it be electronic or even use electrons? Would it be encoded with bits and bytes? It may seem futile to speculate about the past, but it isn't futile to speculate about the future—and speculating about the past can help us choose a different future path.

Or consider this even more outlandish line of inquiry: What future form of computer storage could we imagine being used to store information for, say, interplanetary travel? Would it be a device that performed quantum computing, within the DNA of some organism, or a wetware computer? The answer isn't yet apparent—and it's up to us to decide.

At their best—and perhaps more than any other profession—designers apply their thinking across materials, methods, sectors, and professions, and enable individuals and organizations alike to imagine and deliver change on a global scale. We need not only to think like designers, but to go even further in our thinking. We might even paraphrase Francis Fukuyama and ask not only how we get to Denmark, but how we get beyond it.

The following questions might help us find the way:

- When design stops being human-centric, what is a "short" or "long" time span when we design for humanity, life per se, or for planets?
- How do we care for more people and species, and make the remote feel close and personal?
- How do we shift from human-centered design to life- or planet-centered design?
- Can we expand what is valuable so that the sustainable creation of value becomes the default operating model of both public and private enterprise?
- How do we straddle the material and immaterial world? How do we expand the dimensions for and within which we design?

- How do we expand our view on sectors and break open silos for good?

The six expansions we have proposed all concern areas where even a slight shift in perspective could have a big impact. And although we cannot predict the future, by deploying the six expansions, we can challenge our thinking and create new opportunities for businesses, governments, and individuals alike.

If you wish to start to apply, or try out, the six expansions in practice, you now have two options. Either you can skip straight to the appendix, which provides an overview of the essence of each expansion and some guiding prompts to get you started on challenging your thinking. Or you can go on reading the rest of this chapter to see some ways the expansions could be used to drive innovation in practice. What follows are four examples of what could happen as a consequence of deploying some (if not all) of the six expansions *in combination*. Our examples are partly thought experiments, partly developments already likely to happen...

EXPANDING THE FUTURE OF PRODUCTS

As we've seen, within a very short period of time, we will most likely need to change how we produce and consume things. This will require a major upheaval, but it will also provide a unique opportunity. In particular, we will demand more climate-friendly products—which will require industries that are suitably adapted to this. Biotech—including biodesign—is probably the sector that will grow the most in the near future. In fact, within a few years, biodesign firms may well become some of the world's leading companies, just as Amazon, Google, and Apple once replaced fossil-fuel firms and manufacturing giants. And yet the world's hundred largest design companies are still without biodesign departments.

Now consider the six expansions. We could, for example, apply the *dimensions* expansion and imagine a world in which organic materials replaced digital ones (just as digital once replaced analog). Smart design coupled with heavy biotechnology could spin-off products by manipulating DNA, genetically modifying organisms, or developing computer processors based on biological neurons.

Or consider the *life* expansion. Biodesign includes the development of bacterial processes for coloring clothes (a process that, compared to current chemical processes, requires 99 percent less water). Biodesign includes the development of construction materials made with fungi or algae, as well as the development of sustainable alternatives to meat.

Meanwhile, a world of wetware and biodesign would have legal and regulatory implications and challenges and require public bodies and private enterprises to work together differently—demanding the expansion of our thinking about *sectors* and the respective roles of business and government. After all, designers who have the creativity to develop tomorrow's products are often hamstrung by their lack of access to the kind of expensive laboratory equipment normally reserved for researchers and the largest companies. Companies could set the right framework for designers to collaborate with researchers across sectors. Space should be made for experimentation in order to develop biodesign products that will be attractive to global investors.

Similarly, designing algae-based tableware might seem simple—but not if the products can't get to the market. The legal framework may need updating. For example, in Europe, specific rules determine which materials may be used if the product is intended to come into contact with food. These otherwise sensible restrictions remain unsuitable for "natural" products, such as algae-based plates.

Finally, as opportunities in biotech multiply, we could also expand our thinking about the field in terms of *time* and see it as a ten-year rush to market, a fifty-year challenge, or a hundred-year opportunity.

EXPANDING THE FUTURE OF FLIGHT

Alternatively, consider how expansive thinking might be applied to the future of flying—where it might take us, quite literally. Air travel and related carbon emissions are responsible for more than 2 percent of global emissions (by pre-COVID-19 measures). What are some design ideas that could enhance airplanes to make them more energy-efficient or even replace them?

Deploying the *time* expansion, we can take the long view—and flip back through the calendar to bygone eras and discover discarded and seemingly forgotten technology that is better suited to today's cities and commutes. For instance, almost a century after the Hindenburg disaster, lightweight blimps and airships running on helium are making a comeback. Given the current craze for drones, it is somewhat surprising that their seemingly endless applications haven't been well capitalized. Airships can also be remote controlled and don't suffer from the same excess of energy expenditure that drones carry with them. In fact, airships represent the only form of air transport ready to go carbon neutral instantly.

Although the airship was largely discontinued as a means of passenger transport after the Hindenburg disaster, there is an argument to be made that replacing hydrogen with helium would open many new possibilities for the airship. Even hydrogen could today be safely compartmentalized to avoid a comparable disaster. Where consumers usually choose planes for their speed, slower transport is steadily becoming more popular with the "slow living" movement, nostalgia, and green consciousness. The choice between fast and slow is likely to become more conscious. You might as well book a low-emission train, a plane, or a telecommuting meeting through the same platforms. In fact, lighter-than-air vessels, airships, seem to be a realistic bet for the future and probably the fastest way to zero-emissions flight. To name one, the Airlander 10, from Hybrid Air Vehicle, a new type of hybrid airship, has already been built and is ready for commercial rollout.

Some of the largest wind-turbine manufacturers collaborate on having airships transport their turbines. The biggest wind turbine today, the Haliade-X offshore turbine, has a 220-meter rotor and a 107-meter blade. These are almost impossible to transport on normal roads, and moving from factory to wind farm poses a significant challenge—one that could be solved by the airship.

Without much redesign, airships could also replace cruise ships and even commercial heavy lifting to areas with low infrastructure. You can already preorder an airship cruise to the North Pole. And when gliding slowly over the Atlantic, why not pop into the metaverse a few times to speed things up a bit?

Another long-abandoned option is the ground-effect vehicle (GEV), perhaps better known as the *ekranoplan*: a Soviet-era part plane, part hovercraft. It was designed to attain sustained flight slightly above a level surface—namely, water—using the aerodynamic interaction between the wings and the surface. The energy consumption of the ekranoplan is roughly half that of propeller planes. When looking at the advantages, this could be another technology deemed inappropriate because it was deemed politically unsound. The Soviet-built Lun-class ekranoplan had a range of two thousand kilometers and a potential cruising speed at 550 kilometers per hour, making it a viable fuel-efficient alternative on many routes. Investors already see the potential here. The Sea Wolf Express, a startup, plans to start a commercial ekranoplan route between Helsinki and Tallinn—two capital cities separated by an 80-kilometer stretch of water, the Gulf of Finland. The Iranians also have ekranoplans in their fleet. The benefits are obvious: an ekranoplan is fast and roughly twice as fuel efficient as a plane, and it needs limited infrastructure. It is classified as a boat, but it could easily stand in for planes for long-distance travel where the geography permits. DARPA and Boeing have since been exploring GEVs, and specialized manufacturers in Russia still make them. They lend themselves well to newer vertical liftoff technology. The potential of using ekranoplans even for future intercontinental travel shouldn't be discounted.

We have also seen developments with the designs of electrical vertical takeoff and landing flights (eVTOLs). These have mostly been for solo passengers—and if we're to reduce our carbon emissions in the future, mass transit rather than single-person transportation would be the solution. Compared to fossil-fueled propulsion, the potential for optimal and efficient use of energy with this technology is huge. As a consequence, we might see the return of multipropeller aircraft, as it is much easier to divert electrical power to multiple engines as opposed to the concentrated power use of jet engines. This also opens up the possibility of revisiting other forgotten inventions. A host of rotorcrafts are waiting to be explored and repurposed. For example, the *Fanwing* mixes a fixed wing and a sort of fan. The *cyclorotor* has a type of rotor wing that looks like the ones on a big Mississippi River steamboat. The *Flettner plane* creates lift by using the so-called Magnus effect. The Flettner technology is used today to propel ships, including by Danish ferry operator Scandlines, to save energy.

Other rotorcrafts, like the *autogyro* and the *gyrodyne*, are currently being recycled into eVTOLs. We seem to forget that technologies that didn't succeed failed in a specific context, each with its own culture, economy, and constraints. For every new context, old technology potentially stands a new chance. Who knows, maybe ekranoplans might have been widely used if the technology hadn't originally been spearheaded by a country that collapsed?

The possibilities are endless. Aircraft manufacturers Airbus have formed a design division to look even further into the potential of biomimicry. Experiments are being made using the large wingspan of the albatross as inspiration. With a special technique, the albatross can lock its wings in a position that allows it to glide without flapping its wings for many kilometers.

Such clean-sheet designs are neither the only nor even the easiest ways to mitigate emissions as part of responding to the climate crisis. The COVID-19 pandemic highlighted our need for innovative solutions to long-distance travel, and we need to pursue multiple avenues. Compared to urban transport, which with the right support could be made emission-free fairly quickly, it will be almost a decade before predicted technology improvements (green fuels, batteries, solar-powered planes) allow airplanes to roam the skies without polluting at the levels they do today. And, of course, the future of international air travel depends on the community of international decision makers making the right choices. These choices include a carbon tax and assurances that the political framework is set in place to support a reduction of aviation emissions. But with these assurances, perhaps we'll see long-abandoned aircraft designs revisited, present-day designs enhanced, and new solutions surface. We need to realize that technologies can be abandoned for reasons that cease to exist—reasons that are purely contextual.

EXPANDING THE MOBILE CITY

In the introduction to this book, we said that rapid urbanization is one of the biggest challenges facing humanity today. And, as the human race converges on the city as our main place of living and working, our demand for mobility in cities grows as well. Consider how expansive thinking could change how we get around our fast-growing cities. Urban mobility is affected by how a city is built, its topography, commuting distances, and so on. As such, there is no silver bullet. Experts agree that urban mobility needs to be multi-modal. Today, however, three heavily hyped solutions tend to dominate discussions about how to move large numbers of people around our cities: bike-sharing schemes, electric self-driving cars, and e-scooters.

Bike-sharing programs may be the big countercultural alternative but attempts to scale too fast have gone disastrously wrong—most noticeably in China, where rapid growth has exceeded public demand. Many market entrants failed, and millions of bicycles ended up abandoned, broken, or impounded. As Erdem Ovacik, founder of the Copenhagen-based bike-sharing startup Donkey Republic, told us, the Silicon Valley model of scaling technology at "ten-times growth"—especially hardware—is "too fast and creates a lot of waste." Indeed, if we were to deploy the time and value expansions, we would probably avoid imposing short-term and unsustainable financial business models on our streets—the kind that ends up with huge piles of abandoned and broken bicycles on the street.

Likewise, by expanding our thinking about e-scooters, would we permit so many on our streets? Would we permit them at all? After all, some experts believe e-scooters last just months before breaking down. Others believe they may be a flash in the pan, easily replaced by the next big thing out of Silicon Valley (hoverboards, anyone?). Fortunately, there are signs of mobility companies adopting a more long-term view. At an urban mobility summit in Copenhagen in 2019, Kristian Agerbo, head of public policy at e-scooter firm VOI, conceded that "if these scooters don't last, we don't make any money."

Urban Air Mobility is the last urban air mobility craze with the Silicon Valley mindset: Imagine eVTOL vehicles zooming around from skyscraper to skyscraper. Is this yet another individualistic tech solution without profound thought as to how that would influence society and our cities in terms of energy consumption, noise pollution, social interaction, security, and safety? Certainly, it is likely that cities will go vertical to increase density and make space for the increased number of city dwellers. We will need to figure out how to speed up access to top stories in buildings regardless of if we build "vertiports" or not. Bicycles will need to go vertical too. We are seeing interesting solutions in this respect in the Netherlands: bike lifts and suspended bike lanes. We will need to develop and agree on protocols to safely give robot couriers access within buildings. In China there are ongoing experiments with elevated buses that are designed to arch over the highway traffic. So Vertical Urban Mobility does not necessarily imply Urban Air Mobility.

Nobody knows what the ideal way to get around in cities is yet—not just because of the immense contextual complexity, but also because of the unforeseen speed of change. New types of means of transport—modalities, typologies—are being invented in real time now. It is up to us to make the right choices for our cultural context, for our businesses, our societies, and the planet at large.

By thinking expansively about *time* and *value*, we would be less bullish about electric vehicles, too, in a city context. The car-centric model of urban mobility has failed, and EVs remain unsuitable for use in city centers because they take up as much space as their combustion-engine counterparts. Using ballpark figures, EVs cost around one hundred times as much as any decent bicycle while being 40 percent slower during rush hour, using 2,700 percent more energy, and taking up 600 percent more space on the road.

By thinking *proximity*, we might factor in that in Africa—the continent with the most growth in terms of population and economy—the dominant means of transport are mini-bus-sized share taxis, most notably Kenya's colorful *matatus*. They are extremely cost-, space-, and energy-efficient and highly flexible. The government in Singapore has implemented the concept, and more bus-size sharing concepts are underway. Who says this form of transport will not be the future of motorized urban mobility?

More generally speaking, we could deploy the sectors' expansion and view urban mobility not as a winner-takes-all competition, but as a wicked challenge that can only be solved when the public, private, and civic sectors work together toward the same long-term goals. By thinking expansively about *time*, *value*, and *sectors*, we might better help city planners, regulators, startups, incumbents, and philanthropists break out of their silos and find common cause—and help us adopt metropolitan mobility systems that are human-centered, integrated, holistic, multimodal, and long-term. The point is that narratives around urban mobility and personal transportation options in tomorrow's cities merely reflect hype and demonstrate how getting locked into a particular technological path all too easily obscures or ignores the true design challenge—in this case, space management, sustainability, and quality of life.

Speaking of space management, let's finally consider a use of expansive thinking that was once the preserve of science fiction, but is fast becoming feasible—a voyage to Mars.

EXPANDING THE MOON SHOT

It's become something of a cliché to call for moon shot thinking—and an ironic one too. While the original moon shot is a widely celebrated example of long-term planning, it seems somewhat short-sighted in retrospect. In 1961, as a response to Soviet superiority in the space race, President John F. Kennedy said he wanted to send an American to the moon and safely back before the end of the decade. In theory, the aim should be the benchmark for policy makers: decide on an ambitious target and, with a strong vision, set out to reach it. In hindsight, however, Kennedy's goal was myopic. Some space historians today argue that by focusing all American expertise on one symbolic goal, the big picture got lost. The hope of achieving the manned exploration of Mars was left in the dust, along with the development of more advanced means of propulsion that could have enabled us to colonize remote corners of space.

Six decades on from Kennedy's call for expansive thinking, our collective imagination is once again turning to the Red Planet. Among them is architect Bjarke Ingels, who is developing bold plans for the so-called Mars Science City near Dubai—a space-simulation campus where scientists will work on "humanity's march into space." To many people, of course, making any kind of plan to colonize Mars is lunacy. Our home planet is burning, after all. Our problems lie closer to home.

Ingels prefers to take the long view and says there are at least three reasons why we should endeavor to inhabit the Red Planet. First and foremost, he says, "Life has always been about migrating to places that are difficult to survive and then somehow adapting to that"—first through Darwinian evolution and eventually as we developed intelligence, tools, and technology. "In that sense, it is our destiny to constantly become better at inhabiting more of the world—and traveling through space is just crossing another kind of ocean," Ingels says.

Second, "every time we go to a place that is more difficult to inhabit, we can go back to where we came from with the new tools and inventions that we had to make and apply them wherever we came from." Ingels argues that at least eight of the seventeen UN Sustainable Development Goals deal specifically with the physical environment—"and all of those are very much addressed by the challenges we will face on Mars." In other words, forced to be more efficient at managing natural resources such as water or growing enough food to survive on Mars, we will invent ways to ensure our survival back on Earth.

Finally, Ingels points out that Earth's climate crisis has been caused by our dependence on fossil fuels and carbon emissions.

"On Mars, there are no fossil fuels, because there are no fossils," he says. "So, all energy on Mars will have to be renewable. If we master survival and a sustained human presence on Mars, we already have all the tools and technologies that will allow us to be great custodians of Earth. You could really say that the answer to life on Earth is to be found on Mars."

Stretching your thoughts and efforts to another planet is a result of expansive thinking. In this example from Mars, time and geography—if that term even applies—is expanded. It is expanded for a design purpose beyond the entertainment value of ray guns and macrocephalic gnomes.

Still, if thoughts of the heady Martian atmosphere are getting too much, let's return to the comfort and familiarity of Earth—but briefly consider what might happen if we tried to expand the very expansions we've proposed.

EXPANDING THE EXPANSIONS

Design has huge potential to contribute to positive change, if used strategically as well as ethically. It involves a negotiation between technology, policy, systems, people, and nature to create new futures. A structured design approach can increase the hit rate at the fuzzy front end of innovation in public and private sectors. And design has become incredibly multifaceted, a mechanism for inquiry, for expressing ideas, exploring difficult questions, and addressing global challenges.

In recent years, we've witnessed a global pandemic, continued climate disruptions, Brexit, NASA's discovery of water molecules on the moon, tanking oil prices, and the impeachment of an American president twice. Then there's our growing awareness of racial injustice, countries taking steps toward decarbonizing their industries, a digital transformation of the world's leading companies, and the list goes on. Looking back, it's easy to acknowledge that the world is growing increasingly volatile, uncertain, complex, and ambiguous (VUCA), to paraphrase leadership scholars Warren Bennis and Burt Nanus (and the US Army War College, which first introduced the concept). Acknowledging that comes with an acceptance that current design thinking also needs to adapt to better suit a world that grows more complex by the day—at a pace that is only accelerating.

We need to expand our thinking by merging ideas and thoughts from diverse geographic regions, belief systems, schools of thought, industries, and disciplines. These are in a way our point of departure; they define where we are and what we know and where we might go. From there we can choose our course. Our point of departure is, so to speak, our worldview. We need to expand from somewhere. Only when we have a point of departure can we navigate and expand rapidly, adapting to short-term changes while keeping the long-term challenges and goals in view.

The more cognitive diversity we embrace—in other words, different approaches, experiences, skills, and perceptions of the world—the easier it is to think in an expansive way.

We hope we've shown you not what we think *will* happen in the future but rather *how* a different way of thinking could be applied to start shaping it. We explored six ways to expand our thinking, but we're certain that there are other expansions out there waiting to be discovered and applied. Indeed, we're quite sure there are many other ways of thinking expansively, of thinking beyond design thinking, and stretching the future. Our

APPLYING THE EXPANSIONS

intention has merely been to provide you with the blueprint and the inspiration to pursue whichever path you think is right.

Similarly, we began the book by describing Denmark as a "well-designed" country, and scattered examples of good Danish design throughout the book to demonstrate what we mean by expansive thinking. Though we hope they inspire you, we don't believe they are exceptional. Quite the opposite, in fact. The six expansions are universal in nature and can be used anywhere by anyone. We think they embody a fresh way of thinking, and our hope is that by deploying them you will be able to imagine alternative futures and ways of getting there. It's up to you to find them, of course—but we hope and believe you can.

APPENDIX

EXPANSIONS IN SUMMARY

This appendix provides a brief overview of the essence of each expansion. Additionally, we suggest a number of prompts—or key questions—that can be used to stimulate thinking, group processes, and strategic deliberations in applying the expansions to life and work.

APPENDIX

Expansion	Current state	Expanding to
Time	Short-term (fast-track design processes; quarterly results) Medium term (3- and 5-year strategies and plans) Analytical, linear approaches to strategy Design for shor-term consumption and obsolesence	Ultra-short term (milliseconds) Long term (10–30 years) Extremely long term (300 years plus) Deploying scenarios, science fiction, and design fiction Design for product longevity Design for systems transition
Proximity	Distance between people across gender, ethnicity Distance between people and planet Distance between political and economic elites and populations Designing for exclusion Proximity as closeness to people	Embracing proximity with all humans Empathizing with nature and all living things Reconnecting citizens with political decision makers and administrators Proximity as closeness to nature and planet

Prompts

- What happened in this space historically? When, how, and where did this problem start, and what can we learn from this?
- What is the maximum time horizon for which the innovation or change I am involved with will have an impact—directly or indirectly?
- Is the time horizon so long—and the uncertainty so high—that multiple future outcomes could happen?
- Would our success be highly contingent on the external context? Which "weak signals" of change should we look for?
- How might scenarios or science fiction thinking inspire our work?
- How might I design, commission, or even just purchase something iconic or very high quality so that it might be longer lasting?
- What course of action could we take that would significantly alter the future?

- Which stakeholders do we actually care for?
- Should we expand who or what we care for when we design new solutions?
- Could we look to faraway places for inspiration, increasing geographical proximity?
- What does the increasing connectivity and speed imply for our opportunities to create new value?
- How do we design for "the people not in the room"—for instance, across different socioeconomic groups and gender?
- What would it take to move beyond human-centered to life-centered design?
- What if we must be in proximity to multiple life forms at the same time?

APPENDIX

Expansion	Current state	Expanding to
Life	Today's life span Living human beings Physical "dead" matter as building material Human-centered design Humanity	Ultra-long life spans Digital and artificial (after) life Living organisms as building materials Life-centered design Posthumanity
Value	Linear business models Value creation as transactional Protecting intellectual property rights	Circular economy Value creation as systemic and networked Open source and new forms of IP

APPENDIX

Prompts

- What are the likely consequences of extending the human life span for individuals, business, and society?
- How might we relate to people who continue their lives as digital versions of themselves?
- Could our computers become organic, and how would this change how we relate to them?
- How would new production systems, whether biological or distributed, affect us?
- How would robot-human partnerships influence our productivity, creativity, and capacity for problem-solving?
- Which teams and skills should we cultivate in such a scenario?
- What characterizes our current model of value creation?
- Which parts of our activities extract value, and which parts generate value?
- Would it be possible to shift toward a circular economic model? What would such a shift require?
- What other kinds of value (than purely economic or financial) is generated across the network of actors that our organization is part of?
- Might we visualize the actors, relations, and types of value exchanged in the network in order to find new opportunities for value creation?
- Which new service or product offerings could we introduce in the value-creating system that we are part of, and should we place ourselves differently in the system?
- What would happen if we thought beyond people and organizations to our entire human habitat as we consider the impact we have and the value we can create?

APPENDIX

Expansion	Current state	Expanding to
Dimensions	Traditional physical dimensions Design for Earth Digital (binary) Human intelligence AI for product perfomance	Micro Ultra-large Geo-engineering Design beyond Earth Biological (wetware) Digital *plus* human intelligence AI for creativity
Sectors	Public and private Isolated and separate Traditional means of organizing Legal governance	Public, private, academic, civic Multiple and mutually interconnected sectors Collaborative commons Platforms Cooperatives Agile governance

APPENDIX

Prompts
- What might be the benefits if we radically expanded the scale at which we design—toward the ultra-small or the ultra-large?
- Could we blend physical and digital dimensions, for instance, using VR, and if so, would there be value in this?
- How do we ensure deeply ethical approaches to the way we use the possibilities in AI and data?
- How could we create opportunities for humans and machines to collaborate in meaningful, productive ways?
- What is beyond digital and other habitual ways of working?
- Which silos would bring value to us if they were transgressed?
- How would we break down barriers between sectors?
- How are the increasingly blurred lines between public and private sectors relevant to us?
- Are we aware of the dynamic and entrepreneurial capacity of the public sector and how it might benefit us?
- Could new hybrid forms of public-private collaboration be relevant to explore?

ACKNOWLEDGMENTS

Even though we are both experienced authors, a book is always a demanding project. And it is never accomplished without a lot of great people joining in to help get the job done.

First of all, thanks to James Clasper, brand journalist, for close collaboration from early ideas and research and throughout the writing process. A true pleasure working together. As we proceeded to editing, Katie Dickman at BenBella was a great support, perfectly balancing the delicate line between sound critique and positive encouragement. Thanks also to Matt Holt for support from the outset.

We owe our thanks as well to Esmond Harmsworth of Aevitas Creative for guiding us from the initial pitch to navigating us through to completion of the work.

Further, we are indebted to Danish Design Center research assistants Philip Tørring, Lasse Langstrup Hägerstand, Kamilla Demuth Lund Pedersen, and Helena Mladenovski for diligent and timely work on referencing.

ACKNOWLEDGMENTS

A huge thanks also to Lena Velez Larsen for project management support and scheduling, and the team at Manyone for building the book's flash site. We wish to acknowledge the kind support of Peter Clausen and Per Have and the BMC Foundation, without which this book would not have been possible.

Finally, it goes without saying that writing a book is just an addition to other commitments, which means it cannot avoid influencing family life. Thanks to our families for allowing us the flexibility and space to work on *Expand*. Here's hoping that our ideas and visions will contribute to a better world not only afar, but close to home as well.

REFERENCES

INTRODUCTION

Scott Bronstein and Drew Griffin, "A Fatal Wait: Veterans Languish and Die on a VA Hospital's Secret List," CNN Health, April 23, 2014, https://edition.cnn.com/2014/04/23/health/veterans-dying-health-care-delays/index.html.

Kate Macri, "How the VA's Approach to Human-Centered Design Improved Veterans' Experience," Government CIO Media and Research, June 26, 2020, https://governmentciomedia.com/how-vas-approach-human-centered-design-improved-veterans-experience.

Associated Press, "IG Report: 300,000 Military Veterans Likely Died While Waiting for Healthcare at VA," Military News, September 4, 2015, https://www.military.com/daily-news/2015/09/04/ig-report-300000-veterans-died-while-waiting-health-care-va.html.

Kavi Harshawat and Mary Ann Brody, "New Tool Launches to Improve the Benefits Claim Appeal Process at the VA," US Digital Service, April 21, 2016, https://medium.com/the-u-s-digital-service/new-tool-launches-to-improve-the-benefits-claim-appeals-process-at-the-va-59c2557a4a1c#.3a3gzc76b.

"Isakson, Blumenthal Announce Bill to Modernize VA Appeals Process," The United States Senate Committee on Veterans Affairs, May 2, 2017, https://www.veterans.senate.gov/newsroom/majority-news/isakson-blumenthal-announce-bill-to-modernize-va-appeals-process.

REFERENCES

Allison Arieff, "Designs on the VA," *New York Times*, February 24, 2017, https://www.nytimes.com/2017/02/24/opinion/designs-on-the-va.html.

Against Tech Determinism

Peter Lunenfeld, "The California Design Dominion: Thirteen Propositions," *Los Angeles Review of Books*, December 30, 2019, https://lareviewofbooks.org/article/california-design-dominion-thirteen-propositions/.

Richard Barbrook and Andy Cameron, "The Californian Ideology," *Mute*, September 1, 1995, http://www.metamute.org/editorial/articles/californian-ideology.

John Naughton, "Think the giants of Silicon Valley have your best interests at heart? Think again," *Guardian*, October 21, 2018, https://www.theguardian.com/commentisfree/2018/oct/21/think-the-giants-of-silicon-valley-have-your-best-interestsat-heart-think-again.

Alex Davies, "As Uber Launches Self-Driving in SF, Regulators Shut It Down," *Wired*, December 14, 2016, https://www.wired.com/2016/12/ubers-self-driving-car-ran-red-light-san-francisco/.

Sam Levin and Julia Carrie Wong, "Self-driving Uber Kills Arizona Woman in First Fatal Crash Involving Pedestrian," *Guardian*, March 19, 2018, https://www.theguardian.com/technology/2018/mar/19/uber-self-driving-car-kills-woman-arizona-tempe.

Daniel L. Doctoroff and Will Fleissig, "The Neighbourhood of the Future Starts with Your Ideas," *Toronto Star*, November 1, 2017, https://www.thestar.com/opinion/commentary/2017/11/01/the-neighbourhood-of-the-future-starts-with-your-ideas.html.

Laura Bliss, "Meet the Jane Jacobs of the Smart Cities Age," *Bloomberg*, December 21, 2018, https://www.citylab.com/life/2018/12/bianca-wylie-interview-toronto-quayside-protest-criticism/574477/.

John Naughton, "Don't Believe the Hype: The Media Are Unwittingly Selling Us an AI Fantasy," *Guardian*, January 13, 2019, https://www.theguardian.com/commentisfree/2019/jan/13/dont-believe-the-hype-media-are-selling-us-an-ai-fantasy.

Joichi Ito, "Designing Our Complex Future with Machines," *Journal of Design and Science* (2017), doi:10.21428/8f7503e4.

Jill Lepore, "Our Own Devices—Does Technology Drive History?" *New Yorker*, May 12, 2008, https://www.newyorker.com/magazine/2008/05/12/our-own-devices.

REFERENCES

Adam Satariano, "The World's First Ambassador to the Tech Industry," *New York Times*, September 3, 2019, https://www.nytimes.com/2019/09/03/technology/denmark-tech-ambassador.html.

"The Copenhagen Letter," accessed November 8, 2021, https://copenhagenletter.org.

Nicola Twilley, "When a Virus Is the Cure," *New Yorker*, December 14, 2020, https://www.newyorker.com/magazine/2020/12/21/when-a-virus-is-the-cure.

Six Expansions

Allison Arieff, "Designs on the VA," *New York Times*, February 24, 2017, https://www.nytimes.com/2017/02/24/opinion/designs-on-the-va.html.

Robin McKie, "Nicholas Stern: Cost of Global Warming 'Is Worse Than I Feared,'" *Guardian*, November 6, 2016, https://www.theguardian.com/environment/2016/nov/06/nicholas-stern-climate-change-review-10-years-on-interview-decisive-years-humanity.

DESIGN PAST AND PRESENT

Why Design

Herbert A. Simon, "The Science of Design: Creating the Artificial," *Design Issues* 4, no. 1/2, *Designing the Immaterial Society* (1988): 67–82.

HermanMiller Editors, "Design Q & A: Charles and Ray Eames," *WHY* magazine, accessed November 8, 2021, https://www.hermanmiller.com/stories/why-magazine/design-q-and-a-charles-and-ray-eames/.

David Graeber, *The Utopia of Rules: On Technology, Stupidity, and the Secret Joys of Bureaucracy* (Brooklyn: Melville House, 2013).

Expanding Design Thinking

Eli Meixler, "IBM Is Making Its Design Thinking Available to Clients, Says Its Design Chief," *Fortune*, March 7, 2018, http://fortune.com/2018/03/07/ibm-enterprise-design-thinking/.

Rebecca Linke, "Design Thinking, Explained," MIT, Career Development Office, August 19, 2020, https://cdo.mit.edu/blog/2020/08/19/design-thinking-explained/.

Bruce Nussbaum, "Design Thinking Is a Failed Experiment. So What's Next?" *Fast Company*, April 4, 2011, https://www.fastcompany.com/1663558/design-thinking-is-a-failed-experiment-so-whats-next.

REFERENCES

Helen Walters, "The Real Problems with Design Thinking," *Fast Company*, July 22, 2011, https://helenwalters.com/2011/07/22/the-real-problems-with-design-thinking/.

Mike Monteiro, *Ruined by Design*, (San Francisco: Mule Books, 2019).

Natasha Jen, "Design Thinking Is Bullsh*t," Recorded Talk from 99U Conference, 13:27. Posted by "99U." June 7–9, 2017, https://99u.adobe.com/videos/55967/natasha-jen-design-thinking-is-bullshit.

Roberto Verganti, *Overcrowded: Designing Meaningful Products in a World Awash with Ideas* (Cambridge: The MIT Press, 2017).

Katharine Schwab, "John Maeda: 'In Reality, Design Is Not That Important,'" *Fast Company*, March 15, 2019, https://www.fastcompany.com/90320120/john-maeda-in-reality-design-is-not-that-important.

EXPANSION 1—TIME

Paul Feyerabend, *Against Method* (Brooklyn: Verso, 2010).

Stewart Brand, "The Need for, and the Mechanism by Which, The Long Now Foundation Is Attempting to Encourage Long-Term Thinking," accessed November 8, 2021, The Long Now Foundation, http://longnow.org/about/.

Stewart Brand, "Oxford's Oak Beams and Other Stories of Humans and Trees in Long-Term Partnership," December 31, 2014, https://blog.longnow.org/02014/12/31/humans-and-trees-in-long-term-partnership/, Accessed November 8, 2021.

Expanding by Slowing Down

David Fellah, authors' interview, May 23, 2019.

Bloomberg, "Who Knew You Wanted a New Spicy Snickers Bar? Alibaba Did," *Fortune*, October 24, 2018, http://fortune.com/2018/10/24/alibaba-data-mining-unilever-mars-snickers/.

Paul Polman, authors' notes of speech at Davos about Alibaba testing, January 24, 2018.

Olafur Eliasson, authors' interview, September 17, 2019.

Kevin Kelley, "The 10,000-Year Clock," The Long Now Foundation, accessed November 14, 2021, http://longnow.org/clock/.

Anthony Lydgate, "Jeff Bezos and the Clock That Will Outlast Civilization," *Wired*, September 18, 2018, https://www.wired.com/story/wired25-jeff-bezos-10000-year-clock-civilization/.

"About," The Long Now Foundation, accessed November 14, 2021, http://longnow.org/about/.

REFERENCES

Beatrice Pembroke and Ella Saltmarshe, "The Long Time," Medium—The Long Time Project, October 29, 2018, https://medium.com/@thelongtimeinquiry/the-long-time-3383b43d42ab.

Expanding Through Scenarios

Canada's Horizons, accessed November 14, 2021, https://horizons.gc.ca/en/our-work/learning-materials/foresight-training-manual-module-1-introduction-to-foresight/.

Christian Bason, *Leading Public Sector Innovation: Co-Creating for a Better Society* (Bristol, UK: Policy Press, 2018).

Anne-Mette Termansen, authors' interview, February 18, 2019.

Expanding Through Science Fiction

Megan Willett, "Here Are 15 Sci-Fi Books That Actually Predicted the Future," Eyewitness News, December 6, 2018, https://ewn.co.za/2018/12/06/here-are-15-sci-fi-books-that-actually-predicted-the-future.

Dylan Tweney, "May 25, 1945: Sci-Fi Author Predicts Future by Inventing It," *Wired*, May 25, 2011, https://www.wired.com/2011/05/0525arthur-c-clarke-proposes-geostationary-satellites/.

Andrew Maynard, "Twelve Science Fiction Movies with Something to Say About Ethical Technology Innovation," Medium—Edge of Innovation, September 4, 2018, https://medium.com/edge-of-innovation/12-sci-fi-movies-with-something-to-say-about-emerging-technologies-dd0918c11e02.

Jeremy Lasky, "Science Fiction Thinking: Technology, Magic + Perception—Part 1 of 3," Medium—Predict, August 8, 2018, https://medium.com/predict/science-fiction-thinking-technology-magic-perception-part-1-of-3-f07a68bb43bb.

Ann Rosenberg, *Science Fiction: A Starship for Enterprise Innovation* (Copenhagen: Startup Guide World ApS, 2019).

Expanding Through Design Fiction

Bruce Sterling, "Patently Untrue: Fleshy Defibrillators and Synchronised Baseball Are Changing the Future," *Wired*, November 10, 2013, https://www.wired.co.uk/article/patently-untrue.

Peggy Hollinger, "How Companies Draw on Science Fiction," *Financial Times*, October 2, 2017, https://www.ft.com/content/f603e438-a4ba-11e7-9e4f-7f5e6a7c98a2.

REFERENCES

David Alanyón, "SoftBank's Vision of the Future in 300 Years," *Future Today*, August 11, 2018, https://medium.com/future-today/softbanks-vision-of-the-future-in-300-years-f9a6c54961be.

"Forever Starts Now," Rivian accessed November 15, 2021, https://rivian.com/our-company.

"What We Do," All-Party Parliamentary Group for Future Generations, accessed December 16, 2021, https://www.appgfuturegenerations.com/.

"Committee for the Future," Parliament of Finland, accessed November 14, 2021, https://www.eduskunta.fi/EN/valiokunnat/tulevaisuusvaliokunta/Pages/default.aspx.

"Carbon Neutrality Coalition—Plan of Action," Carbon Neutrality Coalition, accessed October 10, 2021, https://carbon-neutrality.global/plan-of-action/.

"UAE Sets Goal to Be Best in Education and Economy," *Gulf News*, September 27, 2017, https://gulfnews.com/uae/government/uae-sets-goal-to-be-best-in-education-and-economy-1.2097219.

"What Does Xi Jinping's China Dream Mean?" BBC, June 6, 2013, https://www.bbc.com/news/world-asia-china-22726375.

Andrew Miller, "China's Hundred-Year Strategy," *Trumpet*, August 2016, https://www.thetrumpet.com/14006-chinas-hundred-year-strategy.

Træna's Development Plan: Tenk Traena, accessed November 10, 2021, http://www.tenktraena.no.

Moa Bjørnson, "Stay Human" (speech to DOGA conference, Oslo, January 10, 2019), https://doga.no/kalender/dagen2019/.

Expansive Thinking and the Design of Everyday Things

Adam Richardson, *Innovation X: Why a Company's Toughest Problems Are Its Greatest Advantage* (San Francisco: Jossey-Bass, 2010).

Saul Griffiths, quoted in Colin Davies and Monika Parrinder, *Limited Language: Rewriting Design: Responding to a Feedback Culture* (Basel, Switzerland: Birkhäuser, 2010).

Mads Nyvold, "Nej, det er ikke en konspirationsteori. Nogle produkter er bare designet til at gå i stykker før tid," Zetland, March 22, 2017, https://www.zetland.dk/historie/s8qDRzGz-aOMNamWw-074b1.

Statistics Denmark fact about Danes' spending, accessed November 14, 2021, https://www.zetland.dk/historie/s8qDRzGz-aOMNamWw-074b1.

"Average American Owns 29 Cell Phones in a Lifetime," Zetland, accessed November 14, 2021, https://www.zetland.dk/historie/s8qDRzGz-aOMNamWw-074b1.

REFERENCES

Ulla Lunn, "Jomfruøerne—fra fest til katastrofe" POV International, February 9, 2018, https://pov.international/jomfruoerne-fra-fest-til-katastrofe/.

Helena Robertsson, Thomas Zellweger, and Josh Wei-Jun Hsueh, "How the World's Largest Family Businesses Are Proving Their Resilience," EY & St. Gallen, last modified June 2021, https://familybusinessindex.com/#impressum.

"Oak Beams, New College Oxford," Atlas Obscura, last accessed October 10, 2021, https://libanswers.snhu.edu/faq/48009.

EXPANSION 2 — PROXIMITY

Peter Singer, "The Sanctity of Life," *Foreign Policy*, October 20, 2009, https://foreignpolicy.com/2009/10/20/the-sanctity-of-life/.

Kwame Anthony Appiah, *The Lies That Bind: Rethinking Identity, Creed, Country, Color, Class, Culture* (New York: Liveright Publishing Corporation, 2019).

Katharine Brooks, "Massive Glaciers Are Melting in Paris and the City Can't Ignore Them," *HuffPost*, December 10, 2015, https://www.huffingtonpost.com/entry/ice-watch-paris_us_5668af72e4b009377b23bdef.

Cynthia Zarin, "The Artist Who Is Bringing Icebergs to Paris," *New Yorker*, December 5, 2015, https://www.newyorker.com/culture/culture-desk/the-artist-who-is-bringing-icebergs-to-paris.

Olafur Eliasson, authors' interview, September 17, 2019.

Expanding "Proximity" Through Ethics

Janus Friis, "Actics: A Dynamic Ethical Tool of Proximity," unpublished paper, 2004.

Richard van Hooijdonk, "The Future of Holographic Technology Will Flip Almost Every Industry On Its Head," *Richard van Hooijdonk* (blog), https://blog.richardvanhooijdonk.com/en/the-future-of-holographic-technology-will-flip-almost-every-industry-on-its-head/.

Joshua Cooper Ramo, *The Seventh Sense: Power, Fortune, and Survival in the Age of Networks* (New York: Little, Brown & Company, 2016).

"Putting the 'Fab' in Fabrication: Manufacturing in the Digital Age," Imagine, June 15, 2017, https://medium.com/space10-imagine/chapter-1-putting-the-fab-in-fabrication-manufacturing-in-the-digital-age-fc9c7670dc5c.

Expanding Care for (Other) Human Beings

Morten Schwarz Lausten, authors' interview, March 19, 2019.

Giorgia Lupi, "Building Hopes," *Giorgia Lupi* (blog), accessed November 14, 2021, http://giorgialupi.com/building-hopes.

REFERENCES

"Syrian Journey: Choose Your Own Escape Route," BBC, accessed October 10, 2021, https://www.bbc.com/news/world-middle-east-32057601.

"The Refugee Challenge: Can You Break Into Fortress Europe?" *Guardian*, last accessed October 10, 2021, https://www.theguardian.com/global-development/ng-interactive/2014/jan/refugee-choices-interactive.

Expanding—In Order to Collapse the Distance . . .

Christopher Lasch, *The Revolt of the Elites and the Betrayal of Democracy* (New York: W. W. Norton & Co, 1996).

Hilary Cottam, "More Money Will Not Fix Our Broken Welfare State. We Need to Reinvent It," *Guardian*, June 21, 2018, https://www.theguardian.com/commentisfree/2018/jun/21/broken-welfare-state-reinvent-it.

Melissa Benn, "How to Fix the Welfare State," *New Statesman*, June 6, 2018, https://www.newstatesman.com/2018/06/hilary-cottam-radical-help-remake-relationships-revolutionise-welfare-state.

Expanding the Future of Democracy by Including Citizens in Policy Making

Bjarke Ingels, authors' interview, June 21, 2019.

Expanding Proximity to Design for Inclusion

Caroline Criado-Perez, "The Deadly Truth About a World Built for Men—From Stab Vests to Car Crashes," *Guardian*, February 23, 2019, https://www.theguardian.com/lifeandstyle/2019/feb/23/truth-world-built-for-men-car-crashes.

Matthew Cantor, "Nasa Cancels All-Female Spacewalk, Citing Lack of Spacesuit in Right Size," *Guardian*, March 26, 2019, https://www.theguardian.com/science/2019/mar/25/nasa-all-female-spacewalk-canceled-women-spacesuits.

Priya Rao, "Math Is Biased Against Women and the Poor, According to a Former Math Professor," The Cut, *New York* magazine, September 6, 2016, https://www.thecut.com/2016/09/cathy-oneils-weapons-of-math-destruction-math-is-biased.html.

Sigal Samuel, "A New Study Finds a Potential Risk with Self-Driving Cars: Failure to Detect Dark-Skinned Pedestrians," *Vox*, March 6, 2019, https://www.vox.com/future-perfect/2019/3/5/18251924/self-driving-car-racial-bias-study-autonomous-vehicle-dark-skin.

Cathy Keck, "Self-Driving Cars Can't Choose Who to Kill Yet, But People Already Have Lots of Opinions," *Gizmodo*, October 24, 2018, https://gizmodo.com/self-driving-cars-cant-choose-who-to-kill-yet-but-peop-1829984331.

REFERENCES

"What Is Rights of Nature?" Global Alliance for the Rights of Nature, accessed November 14, 2021, https://www.garn.org/about-garn/.

"Great News for Rights of Nature in Ecuador," Global Alliance for the Rights of Nature, accessed November 14, 2021, https://www.garn.org/great-news-for-rights-of-nature-in-ecuador/.

"Design Across Biological Scales," Copenhagen Institute of Interaction Design (CIID), accessed October 10, 2021, http://ciid.dk/education/summer-school/ciid-summer-school-costa-rica-2019/workshops/design_biology/.

Serafina Basciano, "A Wearable That Helps to Understand the Needs of Plants," Springwise, February 5, 2021, https://www.springwise.com/sustainability-innovation/computing-tech/wearable-plant-communication.

Expanding Proximity to Save the Planet

Cynthia Zarin, "The Artist Who Is Bringing Icebergs to Paris," New Yorker, December 5, 2015, https://www.newyorker.com/culture/culture-desk/the-artist-who-is-bringing-icebergs-to-paris.

EXPANSION 3—LIFE

John Muir, *John of the Mountains: The Unpublished Journals of John Muir* (Boston: Houghton Mifflin, 1938), 313.

Meara Sharma, "It's Not Too Late to Stem Climate Change. But We Have No More Time to Waste," *Washington Post*, April 26, 2019, https://www.washingtonpost.com/outlook/its-not-too-late-to-stem-climate-change-but-we-have-no-more-time-to-waste/2019/04/26/8406aa64-525b-11e9-88a1-ed346f0ec94f_story.html.

Jim Conca, "Nature on the Eve of Destruction," *Tri-City Herald*, May 27, 2019, https://www.tri-cityherald.com/opinion/opn-columns-blogs/article230836804.html.

Zack Whittaker, "OpenAI Built a Text Generator So Good, It's Considered Too Dangerous to Release," Techcrunch, February 17, 2019, https://techcrunch.com/2019/02/17/openai-text-generator-dangerous/?guccounter=1.

Expanding Our Life Spans

Adam Gopnik, "Can We Live Longer but Stay Younger?" *New Yorker*, May 13, 2019. https://www.newyorker.com/magazine/2019/05/20/can-we-live-longer-but-stay-younger.

Russell Heimlich, "Baby Boomers Retire," Pew Research Center, December 29, 2010, https://www.pewresearch.org/fact-tank/2010/12/29/baby-boomers-retire/.

REFERENCES

Re:Designing Death and pocket decrees, accessed November 14, 2021, https://www.redeath.org/pocket-sized-funeral-decree/ [website under redesign].

Expanding Real Life into Digital Life

Olivia Solon, "How Close Are We to a Black Mirror-style Digital Afterlife?" *Guardian*, January 9, 2018, https://www.theguardian.com/tv-and-radio/2018/jan/09/how-close-are-we-black-mirror-style-digital-afterlife.

Casey Newton, "Speak, Memory," *Verge*, accessed November 14, 2021, https://www.theverge.com/a/luka-artificial-intelligence-memorial-roman-mazurenko-bot.

Marius Ursache, "The Journey to Digital Immortality," *Marius Ursache* (blog), Medium, October 23, 2015, https://medium.com/@mariusursache/the-journey-to-digital-immortality-33fcbd79949.

"The most daring experience ever made on the Internet," Eter9, accessed November 14, 2021, https://www.eter9.com/.

Doug Bolton, "Russian Billionaire Dmitry Itskov Seeks 'Immortality' by Uploading His Brain to a Computer," *Independent*, March 14, 2016, https://www.independent.co.uk/news/science/dmitry-itskov-2045-initiative-immortality-brain-uploading-a6930416.html.

Expanding Living Materials into New Contexts

Andrew Stewart, "The 'Living Concrete' That Can Heal Itself," CNN, March 7, 2016, https://edition.cnn.com/2015/05/14/tech/bioconcrete-delft-jonkers/.

Stephen Gadd, "Engineering Firm Experimenting with 'Grow Your Own' Walls," February 14, 2019, https://cphpost.dk/?p=109255.

"Grow the Future of Materials," MycoWorks, accessed October 10, 2021, https://www.mycoworks.com/.

"Hi, I'm Eben," Eben Bayer, accessed October 10, 2021, https://www.ebenbayer.com/.

Expanding Life Through Living Materials

"The Promise and Perils of Synthetic Biology," *Economist*, April 4, 2019, https://www.economist.com/leaders/2019/04/04/the-promise-and-perils-of-synthetic-biology.

Mike Scott, "Top Company: Bacteria Maker Is Most Sustainable Corporation of 2019," January 22, 2019, https://www.corporateknights.com/reports/2019-global-100/bacteria-maker-most-sustainable-corporation-of-2019-15481152/.

REFERENCES

Akshat Rathi, "The Revolutionary Technology Pushing Sweden Toward the Seemingly Impossible Goal of Zero Emissions," *Quartz*, June 21, 2017, https://qz.com/1010273/the-algoland-carbon-capture-project-in-sweden-uses-algae-to-help-the-country-reach-zero-emissions/.

Barbara Kantrowitz, "Computers as Mind Readers," *Newsweek*, May 29, 1994, https://www.newsweek.com/computers-mind-readers-188830.

Expanding the Scale of Artificial Life

"Why Cobots?" Universal Robots, accessed November 14, 2021, https://www.universal-robots.com/products/collaborative-robots-cobots-benefits/.

Evan Ackerman and Erico Guizzo, "Consumer Robotics Company Anki Abruptly Shuts Down," *IEEE Spectrum*, April 29, 2019, https://spectrum.ieee.org/automaton/robotics/home-robots/consumer-robotics-company-anki-abruptly-shuts-down.

"PARO Therapeutic Robot," PARO, accessed November 14, 2021, http://www.parorobots.com/.

Amy Harmon, "A Soft Spot for Circuitry," *New York Times*, July 4, 2010, https://www.nytimes.com/2010/07/05/science/05robot.html.

Matt Simon, "Lab-Grown Meat Is Coming, Whether You Like It or Not," *Wired*, February 16, 2010, https://www.wired.com/story/lab-grown-meat/.

Expanding from Human- to Planet-Centered Design

Natsai Audrey Chieza quotes, authors' interview, May 22, 2019.

Sara Burrows, "Trees Talk to Each Other in a Language We Can Learn, Ecologist Claims," Return to Now, February 28, 2018, https://returntonow.net/2018/02/28/trees-talk/.

TED Staff, "Allen Savory's 'How to Fight Desertification and Reverse Climate Change': Criticisms & Updates," June 4, 2018, https://blog.ted.com/allan-savorys-how-to-fight-desertification-and-reverse-climate-change-criticisms-updates/.

HM Treasury, "Final Report—The Economics of Biodiversity: The Dasgupta Review," UK Government, August 20, 2021, https://www.gov.uk/government/publications/final-report-the-economics-of-biodiversity-the-dasgupta-review.

"Design for Planet, MA Programme," Designskolen Kolding Design for Planet, https://www.designskolenkolding.dk/en/design-planet.

REFERENCES

Expanding from Humanity to Posthumanity
Sean Captain, "We Don't Always Know What AI Is Thinking—And That Can Be Scary," *Fast Company*, November 15, 2016, https://www.fastcompany.com/3064368/we-dont-always-know-what-ai-is-thinking-and-that-can-be-scary.

Adam Greenfield, *Radical Technologies: The Design of Everyday Life* (Brooklyn: Verso, 2017).

Paola Antonelli, "About," *Broken Nature*, accessed November 14, 2021, http://www.brokennature.org/about/xxii-triennale/.

Mikel Jaso, "Striking a Balance Between Fear and Hope on Climate Change," *New York Times*, April 15, 2019, https://www.nytimes.com/2019/04/15/books/review/bill-mckibben-falter.html.

EXPANSION 4—VALUE
Arthur Huang, authors' interview, January 24, 2019.

Expanding in the Anthropocene
Jonathan Watts, "We Have 12 Years to Limit Climate Change Catastrophe, Warns UN," *Guardian*, October 8, 2018, https://www.theguardian.com/environment/2018/oct/08/global-warming-must-not-exceed-15c-warns-landmark-un-report.

Chelsea Harvey, "Climate Change Is Becoming a Top Threat to Biodiversity," *E&E News*, March 28, 2018, https://www.scientificamerican.com/article/climate-change-is-becoming-a-top-threat-to-biodiversity/.

Homi Kharas, "The Unprecedented Expansion of the Global Middle Class," The Brookings Institution, February 28, 2017, https://www.brookings.edu/research/the-unprecedented-expansion-of-the-global-middle-class-2/.

Doughnut Economics Action Lab, "1. Change the Goal—1/7 Doughnut Economics," YouTube video, April 2, 2017, https://www.youtube.com/watch?v=Mkg2XMTWV4g.

Expanding the Circular Economy
"What Is a Circular Economy?", Ellen MacArthur Foundation, https://www.ellenmacarthurfoundation.org/circular-economy/concept.

"Sustainability Services," Accenture, accessed November 14, 2021, https://www.accenture.com/us-en/sustainability-index.

Matthew McGuiness, "The Circular Economy Could Unlock $4.5 Trillion of Economic Growth, Finds New Book by Accenture," Accenture, September 28, 2015, https://newsroom.accenture.com/news/the-circular-economy-could

REFERENCES

-unlock-4-5-trillion-of-economic-growth-finds-new-book-by-accenture.htm.

James Vincent, "This Adidas Sneaker Made from Recycled Ocean Waste Is Going On Sale This Month," *Verge*, November 4, 2016, https://www.theverge.com/2016/11/4/13518784/this-adidas-sneaker-made-from-recycled-ocean-waste-is-going-on-sale-this-month.

Flemming Besenbacher, authors' interview, April 10, 2019.

"IKEA receives circular economy award," Climate Action, January 29, 2018, accessed November 14, 2021, https://www.climateaction.org/news/ikea-receives-circular-economy-award.

Arthur Huang, authors' interview, January 24, 2019.

Expanding to Value-Creating Networks

Rafael Ramírez and Ulf Mannervik, *Strategy for a Networked World* (London: Imperial College Press, 2016).

"Cradle to Grave Design Paradigm," Method, accessed November 14, 2021, https://methodrecycling.com/world/journal/what-the-nz-circular-summit-means-for-the-future-of-design.

Adlai Stevenson quote, "We Travel Together," *New York Times*, June 2, 2017, https://www.nytimes.com/2017/06/02/opinion/climate-adlai-stevenson-trump.html.

Expanding How Value Is Protected

Jens Martin Skibsted, "Innovative Intellectual Property," Medium, July 23, 2020, https://medium.com/ogojiii/ip-2d1ebd4e0e17.

Arthur Huang, authors' interview, January 24, 2019.

EXPANSION 5—DIMENSIONS

Sam Byford, "AlphaGo Beats Lee Se-dol Again to Take Google DeepMind Challenge Series," *Verge*, March 12, 2016, https://www.theverge.com/2016/3/12/11210650/alphago-deepmind-go-match-3-result.

Chris Baraniuk, "The Cyborg Chess Players That Can't Be Beaten," BBC, December 14, 2015, http://www.bbc.com/future/story/20151201-the-cyborg-chess-players-that-cant-be-beaten.

Wikipedia, s.v. "advanced chess," accessed August 21, 2021, https://en.wikipedia.org/wiki/Advanced_Chess.

REFERENCES

Expanding the Dimensions of Design

"About," Otherlab, accessed November 14, 2021, https://www.otherlab.com/home/#we-are.

"A New Type of Fabric," Skyscrape, accessed November 14, 2021, http://materialcomforts.com/.

Brent Ridley, "From Idea to Invention," Medium, May 25, 2016, https://medium.com/@otherbrent/from-idea-to-invention-43af3bc550ff.

Holly Cave, "The Nanotechnology in Your Clothes," *Guardian*, February 4, 2014, https://www.theguardian.com/science/small-world/2014/feb/14/nanotechnology-clothes-nanoparticles.

Mark Crawford, "10 Ways Nanotechnology Impacts Our Lives," American Society of Mechanical Engineers, March 1, 2016, https://www.asme.org/engineering-topics/articles/technology-and-society/10-ways-nanotechnology-impacts-lives.

"How Does District Heating Work?" Danfoss, https://www.danfoss.com/en/about-danfoss/our-businesses/heating/knowledge-center/heating-school/how-does-district-heating-work/.

"Use of District Heating Equals Higher Energy Efficiency and Individual Comfort," Danfoss, https://www.danfoss.com/en/markets/district-energy/dhs/district-heating/#tab-overview.

"District Heating," State of Green, accessed November 14, 2021, https://stateofgreen.com/en/sectors/district-energy/district-heating/.

"2019 Winner," Danish Design Award, accessed November 14, 2021, https://danishdesignaward.com/en/nominee/energylab-nordhavn/.

Jens Martin Skibsted, "Here's How a Smart Water Grid Could Stop Global Warming," World Economic Forum, January 25, 2018, https://www.weforum.org/agenda/2018/01/heres-how-a-smart-water-grid-could-stop-global-warming/.

Expanding into Geo-Engineering

Jonathan Watts, "Geoengineering May Be Used to Combat Global Warming, Experts Say," *Guardian*, October 8, 2018, https://www.theguardian.com/environment/2018/oct/08/geoengineering-global-warming-ipcc.

Alexis Madrigal, "Q&A: Geoengineering Is 'a Bad Idea Whose Time Has Come,'" *Wired*, March 23, 2010, https://www.wired.com/2010/03/hacktheplanet-qa/.

Jonathan Watts, "Cities Fall Out over Cloud," *Guardian*, July 15, 2004, https://www.theguardian.com/environment/2004/jul/15/china.weather.

REFERENCES

Jonathan Watts, "China's Largest Cloud Seeding Assault Aims to Stop Rain on the National Parade," *Guardian*, September 23, 2009, https://www.theguardian.com/environment/2009/sep/23/china-cloud-seeding.

Jessica Mairs, "Bjake Ingels Proposes Mars Simulation City for Dubai in Race for Space Colonisation," *Dezeen*, September 28, 2017, https://www.dezeen.com/2017/09/28/bjarke-ingels-mars-science-city-space-exploration-dubai-united-arab-emirates/.

Bill Steigerwald and Nancy Jones, "Mars Terraforming Not Possible Using Present-Day Technology," NASA Official, July 30, 2018, https://www.nasa.gov/press-release/goddard/2018/mars-terraforming.

Expanding into Digital

Nick Statt, "Apple Hires Legendary Designer Marc Newson Ahead of 'iWatch' Launch," CNET, September 5, 2014, https://www.cnet.com/news/apple-hires-legendary-designer-marc-newson-ahead-of-iwatch-launch/.

Expanding into VR

Thomas Musca, "Why Henning Larsen Architects Believe That VR Is 'a Gift for the Future of Architecture,'" *ArchDaily*, August 19, 2017, https://www.archdaily.com/876881/why-henning-larsen-architects-believe-that-vr-is-a-gift-for-the-future-of-architecture/.

"10 Best Augmented Reality Apps in the Present Marketplaces," Technostacks, November 1, 2018, https://technostacks.com/blog/best-augmented-reality-apps/.

Jeremy White, "IKEA's Revamped AR App Lets You Design Entire Rooms," *Wired*, April 20, 2021, https://www.wired.com/story/ikea-revamped-ar-app-design-entire-rooms/.

Patrick Tucker, "Tomorrow Soldier: How The Military Is Altering the Limits of Human Performance," *Defense One*, July 12, 2017, https://www.defenseone.com/technology/2017/07/tomorrow-soldier-how-military-altering-limits-human-performance/139374/.

Oscar Raymundo, "Tim Cook: Augmented Reality Will Be an Essential Part of Your Daily Life, Like the iPhone," *Macworld*, October 3, 2016, https://www.macworld.com/article/3126607/tim-cook-augmented-reality-will-be-an-essential-part-of-your-daily-life-like-the-iphone.html.

Expanding into Societal Change

Lily Kuo, "China Bans 23m from Buying Travel Tickets as Part of 'Social Credit' System," *Guardian*, March 1, 2019, https://www.theguardian.com/world/2019/mar/01/china-bans-23m-discredited-citizens-from-buying-travel-tickets-social-credit-system.

Kelsey Munro, "China's Social Credit System 'Could Interfere in Other Nations' Sovereignty,'" *Guardian*, June 27, 2018, https://www.theguardian.com/world/2018/jun/28/chinas-social-credit-system-could-interfere-in-other-nations-sovereignty.

Simina Mistreanu, "Life Inside China's Social Credit Laboratory," *Foreign Policy*, April 3, 2018, https://foreignpolicy.com/2018/04/03/life-inside-chinas-social-credit-laboratory/.

Neda Ulaby, "Dan Ariely Takes on 'Irrational' Economic Impulses," NPR, March 31, 2008, https://www.npr.org/2008/03/31/89233955/dan-ariely-takes-on-irrational-economic-impulses.

"Will AI Help Us Become Better Architects?" GXN, accessed November 14, 2021, https://gxn.3xn.com/project/can-integrate-ai-architecture/.

"AI—Introducing the First Chair Created with Artificial Intelligence," Starck, January 24, 2020, https://www.starck.com/a-i-introducing-the-first-chair-created-with-artificial-intelligence-p3801.

Expanding into Tomorrowland

James Vincent, "Former Go Champion Beaten by DeepMind Retires After Declaring AI Invincible," *Verge*, November 27, 2019, https://www.theverge.com/2019/11/27/20985260/ai-go-alphago-lee-se-dol-retired-deepmind-defeat.

Melanie Lefkowitz, "Chess Engine Sacrifice Mastery to Mimic Human Play," *Cornell Chronicle*, January 25, 2021, https://news.cornell.edu/stories/2021/01/chess-engine-sacrifices-mastery-mimic-human-play.

Brian Turner, "Exploring the Weird World of Exoplanets," TechRadar, March 28, 2019, https://www.techradar.com/news/the-weird-world-of-exoplanets.

EXPANSION 6—SECTORS

Mariana Mazzucato, *The Entrepreneurial State* (London: Demos, 2011).

"Sønderborg Castle," Visit Nordic, accessed November 14, 2021, https://www.visitnordic.com/en/attraction/sonderborg-castle.

"Velkommen til Sønderborg," *Sønderborg*, accessed November 14, 2021, http://sonderborg.dk/da/byens-havn/masterplan.

REFERENCES

"Project Zero Sønderborg," accessed November 14, 2021, http://brightgreenbusiness.com/.

"Knowledgebase," Danish Architecture Center, accessed September 2021, https://arcspace.com/feature/sonderborg-harbor-masterplan/.

Andrew Simms, "Sønderborg: The Little-Known Danish Town with a Zero Carbon Master Plan," *Guardian*, October 22, 2015, https://www.theguardian.com/sustainable-business/2015/oct/22/denmark-sonderborg-danish-town-trying-to-be-carbon-neutral.

"Danfoss Fills Alsik with Energy Effective Technology," Bitten & Mads Clausens Fond, September 3, 2018, http://www.bmcfond.com/news/danfoss-fills-alsik-with-energy-effective-technology/.

"Deloitte at WEF Davos 2020," Deloitte, accessed November 14, 2021, https://www2.deloitte.com/global/en/pages/about-deloitte/articles/davos-insights.html.

"What You Need to Know About Impact Investing," Global Impact Investing Network (GIIN), accessed November 8, 2021, https://thegiin.org/impact-investing/need-to-know/#how-big-is-the-impact-investing-market.

"Sustainable Investing Basics," The Forum for Sustainable and Responsible Investment, September 21, 2021, https://www.ussif.org/sribasics.

Kelly LaVigne, "Investing with a Conscience: The Rise of ESG and What Really Matters to Investors," *Kiplinger*, June 17 2019, https://news.yahoo.com/investing-conscience-rise-esg-really-113621437.html.

"StepJockey: Health App Developer Wins £600,000 Private Investment," Government of United Kingdom, September 21, 2015, https://www.gov.uk/government/case-studies/stepjockey-health-app-developer-wins-600000-private-investment.

Expanding into the Entrepreneurial State

Mariana Mazzucato, *The Entrepreneurial State* (London: Demos, 2011).

"Professor Mazzucato's 'Missions' at the Core of Ambitious New €100bn EU Proposal," Institute for Innovation and Public Purpose, accessed November 14, 2021, https://www.ucl.ac.uk/bartlett/public-purpose/news/2018/jun/professor-mazzucatos-missions-core-ambitious-new-eu100bn-eu-proposal.

"Horizon Europe," European Commission, accessed November 14, 2021, https://ec.europa.eu/info/horizon-europe-next-research-and-innovation-framework-programme_en.

"EIT at a Glance," European Institute of Innovation and Technology, accessed November 14, 2021, https://eit.europa.eu/who-we-are/eit-glance.

REFERENCES

Daria Tataj quote, email correspondence with authors, August 1, 2019.
UNDP Accelerator Labs, accessed November 14, 2021, https://acceleratorlabs.undp.org.
Silvia Morimoto, "It Takes More Than Two to Tango," UNDP, June 11, 2018, https://www.undp.org/content/undp/en/home/blog/2018/It-takes-more-than-two-to-tango.html.
"Public Sector Innovation Facets," Observatory of Public Sector Innovation, accessed November 14, 2021, https://oecd-opsi.org/projects/innovation-facets/.

Expanding the (Even More) Entrepreneurial State

Jerry Hirsch, "Elon Musk's Growing Empire Is Fueled by 4.9 Billion in Government Subsidies," *Los Angeles Times*, May 30, 2015, https://www.latimes.com/business/la-fi-hy-musk-subsidies-20150531-story.html.
Nigel Cameron, "The Government Agency That Made Silicon Valley," *UnHeard*, June 18, 2021, https://unherd.com/2018/06/government-agency-made-silicon-valley/.
Mariana Mazzucato, *The Entrepreneurial State* (London: Demos, 2011).
Bloomberg news editors, "Denmark to End All Renewable Energy Subsidies," Renewable Energy World, accessed November 14, 2021, https://www.renewableenergyworld.com/wind-power/denmark-to-end-all-renewable-energy-subsidies/#gref.
Hamza Bendemra, "NASA's Reliance on Outsourcing Launches Causes a Dilemma for the Space Agency," *Conversation*, July 3, 2021, https://theconversation.com/nasas-reliance-on-outsourcing-launches-causes-a-dilemma-for-the-space-agency-44013.
Wikipedia, s.v. "List of Falcon 9 and Falcon Heavy launches," accessed November 14, 2021, https://en.wikipedia.org/wiki/List_of_Falcon_9_and_Falcon_Heavy_launches.
Andrew Liptak, "NASA Is Hoping to Hand International Space Station Over to Commercial Entity in the Next Decade," *Verge*, August 21, 2016, https://www.theverge.com/2016/8/21/12574300/nasa-international-space-station-commercial-entity-next-decade.
Lucy Ingham, "Privatising the International Space Station Is the Start of the First City in Space," Factor, accessed November 14, 2021, https://www.factor-tech.com/feature/privatising-the-international-space-station-is-the-start-of-the-first-city-in-space/.

REFERENCES

Expanding the Power of Philanthropy

Rupert Neate, "Rockefeller Family Charity to Withdraw All Investments in Fossil Fuel Companies," *Guardian*, March 23, 2016, https://www.theguardian.com/environment/2016/mar/23/rockefeller-fund-divestment-fossil-fuel-companies-oil-coal-climate-change.

Jillian Ambrose, "World's Biggest Sovereign Wealth Fund to Ditch Fossil Fuels," *Guardian*, June 12, 2019, https://www.theguardian.com/business/2019/jun/12/worlds-biggest-sovereign-wealth-fund-to-ditch-fossil-fuels.

Helena Robertsson, Thomas Zellweger, and Josh Wei-Jun Hsueh, "The World's Largest Family Businesses Are Proving Their Resilience," Family Business Index, June 1, 2021, http://familybusinessindex.com.

"About ALU," African Leadership University, accessed November 14, 2021, https://www.alueducation.com/about/.

Anna Mckie, "'Harvard of Africa' Now 'Not an Academic Institution,'" *Times Higher Education*, July 3, 2018, https://www.timeshighereducation.com/news/harvard-africa-now-not-academic-institution.

Yinka Adegoke, "African Leadership University Raises $30 Million to Help Reinvent Graduate Education," *Quartz Africa*, January 4, 2019, https://qz.com/africa/1515015/african-leadership-university-raises-30-million-series-b/.

Fred Swaniker, authors' interview, May 2019.

"Our Mission," World Economic Forum, accessed November 14, 2021, https://www.weforum.org/about/world-economic-forum.

Klaus Schwab, email correspondence with authors, August 15, 2019.

Christian Bason, "Da verdenseliten kastede sig over Faaborg-Midtfyn," *Mandag Morgen*, January 31, 2019, https://www.mm.dk/artikel/da-verdenseliten-kastede-sig-over-faaborg-midtfyn.

Expanding—or Imploding—the Silo

Jeremy Rifkin, "The End of the Capitalist Era, and What Comes Next," *HuffPost*, December 6, 2017, https://www.huffpost.com/entry/collaborative-commons-zero-marginal-cost-society_b_5064767.

"Global Future Council on Agile Governance," World Economic Forum, accessed November 14, 2021, https://www.weforum.org/communities/global-future-council-on-agile-governance.

Wikipedia, s.v. "Mondragon Corporation," accessed October 5, 2021, https://en.wikipedia.org/wiki/Mondragon_Corporation#Business_culture.

REFERENCES

David Colander and Roland Kupers, *Complexity and the Art of Public Policy: Solving Society's Problems from the Bottom Up* (Princeton, NJ: Princeton University Press, 2016).

Martin Stewart-Weeks "Can We Declare a Covid Dividend?," Public Purpose, April 3, 2020, https://publicpurpose.com.au/can-we-declare-a-covid-dividend/.

"About," Impact the Future, accessed November 14, 2021, https://www.impactthefuture.co/about.

APPLYING THE EXPANSIONS

Olafur Eliasson, authors' interview, September 17, 2021.

Gina Cronin, "Quipu: The Ancient Computer of the Inca Civilization," *Peru for Less*, November 30, 2020, https://www.peruforless.com/blog/quipu/.

Expanding the Future of Flight

Jens Martin Skibsted, "The Return of the Blimp," World Economic Forum, August 17, 2016, https://www.weforum.org/agenda/2016/08/rethinking-off-grid-delivery-in-africa.

Jens Martin Skibsted, "Tech Thinking Has Become Too Limiting—It Must Be Disrupted," World Economic Forum, January 22, 2019, https://www.weforum.org/agenda/2019/01/tech-thinking-has-become-too-limiting-it-must-be-disrupted.

"Vind i rotorsejlet," Scandlines, accessed November 15, 2021, https://www.scandlines.dk/om-os/vores-gronne-agenda/vind-i-rotorsejlet/.

Expanding the Mobile City

Allan Taylor, "Bike Share Oversupply in China: Huge Piles of Abandoned and Broken Bicycles," *Atlantic*, March 22, 2018, https://www.theatlantic.com/photo/2018/03/bike-share-oversupply-in-china-huge-piles-of-abandoned-and-broken-bicycles/556268/.

Erdem Ovacik, "Urban Mobility Summit at Techfestival 2019—Recap," Donkey Republic. October 7, 2019, https://www.donkey.bike/urban-mobility-summit-at-techfestival-2019-recap/.

Micah Tool, "Superpedestrian's Self-Repairing Electric Scooter Is Exactly What Lime and Bird Need," *Electrek*, December 4, 2018, https://electrek.co/2018/12/04/superpedestrian-electric-scooter/.

Micah Tool, "The Future of Electric Scooter Sharing Companies—4 Scenarios 4 Years from Now," *Pontoon-e*, December 26, 2018, https://pontoon-e.com/

REFERENCES

the-future-of-electric-scooter-sharing-companies-4-scenarios-4-years-from-now/.

Kristian Agerbo, "Urban Mobility Summit at Techfestival 2019—Recap," October 7, 2019, https://www.donkey.bike/urban-mobility-summit-at-techfestival-2019-recap/.

Expanding the Moon Shot

Jessica Mairs, "Bjarke Ingels Proposes Mars Simulation City for Dubai in Race for Space Colonisation," *Dezeen*, September 28, 2017, https://www.dezeen.com/2017/09/28/bjarke-ingels-mars-science-city-space-exploration-dubai-united-arab-emirates/.

Bjarke Ingels, authors' interview, June 21, 2019.

LIST OF INTERVIEWEES

Flemming Besenbacher, chairman, Carlsberg
Moa Bjørnson, development manager, Træna, Norway
Natsai Audrey Chieza, founder and CEO, Faber Futures
Olafur Eliasson, artist, Studio Olafur Eliasson
David Fellah, co-CEO and founder, Manyone
Arthur Huang, architect and structural engineer, founder of Miniwiz
Bjarke Ingels, architect, founding partner, BIG
Morten Schwarz Lausten, senior consultant, Red Cross, Denmark
Fred Swaniker, chairman and cofounder, African Leadership Academy
Daria Tataj, founder and CEO, Tataj Innovation
Anne-Mette Termansen, head of division, Copenhagen Capital Region

INDEX

A

Accenture, 112, 117
Accurat, 69
actics system, 60–61, 76
A/D/A model, 28
Adams, Douglas, 45, 175
aerosols, 139
aesthetic sustainability, 118
Africa, 62, 126, 167–168
African Leadership University (ALU), 167–168
African Leadership X (ALX), 168
Agerbo, Kristian, 185
AI. *see* artificial intelligence
air travel, 180–184
Alexa digital voice assistant, 47, 141
algae-based tableware, 179–180
Algoland project, 96
algorithms, 79–80, 148
Alibaba, 35–36
Allianz Life, 155

All-Party Parliamentary Group on Future Generations, 50
Alphabet, 11, 123. *See also* Google
AlphaGo, 147
Alsik conference center, 152
ALU (African Leadership University), 167–168
ALX (African Leadership X), 168
Amazon, 36, 47, 49, 123, 179
anthropocene epoch, 87–89, 110–111
Antonelli, Paola, 105
AP Møller/Mærsk Group, 53
appellation systems, 127
Appiah, Kwame Anthony, 57
Apple, 123, 141, 144, 179
AR (augmented reality), 142–144
Arieff, Allison, 3–4
Arla Foods, 171
ARPANET, 161
artificial intelligence (AI), 146–149
    AI-assisted design, 146–147
    in competitive games, 147–148

INDEX

designing, 132
digital afterlifes with, 91
human intelligence and, 129–130
implementation of, 11–12
and machine learning, 77
posthumanity, 104–105
science fiction and, 46–47
artificial life, 97–100
Arup, 49
Atlas Obscura (website), 55
augmented reality (AR), 142–144
autonomous vehicles (AV), 10–11

B

bacteria, in biodesign, 95–96, 179
bacteriophages, 14
Bafokeng Community Model, 165–166
Bang & Olufsen, 34
Barbrook, Richard, 9–10
Barcelona, Spain, 66
Bason, Christian, 6–7, 39, 73
BBC, 69, 143
Bennis, Warren, 190
Besenbacher, Flemming, 113–114
Bestseller Foundation, 168
Bezos, Jeff, 36–37, 49, 164
bias, in algorithms, 79
Big Tech, 9–13, 70
bike-sharing programs, 184–185
BioCollar, 82
bioconcrete, 92–93
biodesign, 27, 178–180
bioengineering, 149
biotechnology, 178–180
Bitten and Mads Clausen Foundation, 152
Bjarne Rasmussen Holding, 152

Bjørnson, Moa, 51
Bleecker, Julian, 47
blockchain technologies, 105
Blumenthal, Richard, 2
Boeing, 182
Bolivia, 81
Branson, Richard, 164
Braungart, Michael, 122
Brin, Sergei, 161–162
Broken Nature (exhibition), 105
Brookings Institute, 111
Brooks, Sarah, 2, 22
Brown, Tim, 28
Burkina Faso, 126
Burton, Tim, 87

C

Cameron, Andy, 9–10
Cameron, Nigel, 162
Campbell, Naomi, 175
Carbon Neutrality Coalition Plan of Action, 50
Carlsberg Group, 113, 167
Chieza, Natsai Audrey, 101
China, 16, 50, 139, 144–145, 184–185
Chr. Hansen (company), 95
CIID (Copenhagen Institute of Interaction Design), 81–82
circular design, 115–118
circular economy, 112–123
creation value in, 114–115
at geopolitical levels, 119–120
sustainable products in, 112–114
value-creating networks in, 120–123
citizen-centered policy, 73
Civilizations app, 143
Civil Service College, 38

INDEX

Clarke, Arthur C., 45
Clausen, Mads, 134
climate change movement, 36, 54, 58–59, 62–63, 83–85, 110–111, 131, 166
climate crisis, 75, 188–189
CNC machines, 65, 66
cobots, 98
cognitive diversity, 189–191
Colander, David, 172
Collaborative Commons, 170–171
collaborative design, 36, 39
Committee for the Future (Finland), 50
Constitute Project, 74
Cook, Tim, 144
cooperative design, 5
COP26 climate summit, 62
Copenhagen, Denmark, 8–9, 39–40, 57–58, 67–68, 135, 154
Copenhagen Institute of Interaction Design (CIID), 81–82
Corporate Knights, 95
Cottam, Hilary, 71–72
COVID-19 pandemic, 84, 183
cradle-to-cradle movement, 122
Cramton, Steve, 130, 140
Creative Commons, 124, 125
Criado Perez, Caroline, 77–78
cross-sector collaboration, 157–162, 168–169
cultural biases, 15–17, 126–127

D

Danfoss, 53, 134–136, 138, 152–153, 167
Danish design, 6–9
Danish Design Centre, 40
Danish Management Society, 154

DARPA (Defense Advanced Research Projects Agency), 161–162, 182
Dasgupta, Partha, 102–103
Davos, Switzerland, 114, 117
death, coping with, 89–90
decision-making, 38–40, 72–75
Deep Blue supercomputer, 129–130
Deep Mind, 147
Defense Advanced Research Projects Agency (DARPA), 161–162, 182
Defense One (website), 143
Degerhamn cement factory (Sweden), 96
Deleuze, Gilles, 151
Delft University of Technology, 93
delivery, designing product, 27
Deloitte, 154, 156
democracy, 72–75
Democracy Earth, 74–75
Demos (think tank), 157
Denmark, 7–9, 52–54, 93, 103, 135, 151–152, 154, 162, 167, 169–170, 174
Department of Health (United Kingdom), 156
design fiction, 47–51
Design for the Planet Program, 103
Designit, 34
design thinking, 1–23, 25–32
  in Danish design, 6–9
  expansive thinking for, 17–18
  flaws in, 27–32
  history of, 26
  in human-centered design, 4–6
  present factors for, 27
  six expansions for, 18–22
  in technological determinism, 9–17
  thinking about, 32

at VA, 1–4, 22
Deutsch, David, 129
Dezeen, 139–140
Diabaté, Issa, 63
digital afterlife, 91
digital development, 34–35
digital ethics, 150
digital fabrication, 65–66
digital lives, 90–92
digital proximity, 67–70
digital technology, 90–92, 140–141
dimensional expansion, 129–150
    about, 20–21
    applying, 198–199
    digital technology for, 140–141
    geo-engineering for, 138–140
    and human capabilities, 129–132
    in metaverses, 147–150
    for product design, 133–138
    societal change for, 144–147
    virtual reality for, 142–144
distance, proximity expansion and, 57–59
district heating, 134–135
Donkey Republic, 184
Drybath Gel, 63
DTU, 93
durable products, 51, 115

E

Eames, Charles, 4
Eat Just, 99
Economic and Social Committee (EU), 52
Economist, 95
Ecovative, 93–94
Ecuador, 81
Eisenhower, Dwight D., 161
EIT (European Institute of Innovation and Technology), 158–159
ekranoplan, 181–182
electrical vertical takeoff and landing flights (eVTOLs), 182–183
Eliasson, Olafur, 36, 57–59, 83, 102, 176
Eliava, George, 14
Eliava Institute, 14
Ellen MacArthur Foundation, 116
empathy, 67, 81–83
"The Enemy" (VR installation), 69
energy efficiency, 133–138
Energy-Lab Nordhavn, 135
energy self-sufficiency, 162
England, 115
Engtoft Larsen, Anne Marie, 13
entrepreneurial states, 157–162
entrepreneurship, 172–173
Ernst & Young (EY), 53
e-scooters, 185
Eter9, 91
Eterni.me, 91
ethical design, 77
ethics, 59–62, 150
ethnicity bias, 79–80
European Institute of Innovation and Technology (EIT), 158–159
European Union (EU), 52, 119, 120, 158, 173
eVTOLs (electrical vertical takeoff and landing flights), 182–183
expansive thinking. *see also specific expansions by name*
    for air travel, 180–184
    cognitive diversity for, 189–191
    deploying, 175–178

INDEX

for design thinking, 17–18
premise of, 5
for product design, 51–55, 178–180
for space travel, 187–189
for urban mobility, 184–187
ExxonMobil, 165
EY (Ernst & Young), 53

F

Faaborg-Midtfyn, Denmark, 169–170
Fab City (Barcelona, Spain), 66
Facebook, 26, 79, 104, 123, 149
Fast Company, 104
Fellah, David, 34–35
Feyerabend, Paul K., 33, 176
Finland, 50
Finless Foods, 99
Ford Motor Company, 49
Fortune, 28
Foster + Partners, 140
France, 14, 69
Fridays for Future movement, 84
Friis, Janus, 60
Fukuyama, Francis, 7, 177
Fuller, R. Buckminster, 107
Furlenco, 122

G

gamification, 69–70
Garland, Alex, 105
Gehry, Frank, 151
gender bias, 77–78
General Electric, 29
geo-engineering, 138–140
geography, proximity and, 62–64
geopolitics, 119–120
Georgia, Soviet republic of, 14

Georgia Institute of Technology, 79
GEVs (ground-effect vehicles), 181–182
GIIN (Global Impact Investing Network), 155
Glasgow, Scotland, 62
Global Alliance for the Rights of Nature, 80–81
Global Family Business Index, 53
Global Future Council, 169–170, 171
Global Impact Investing Network (GIIN), 155
global warming, 138. *see also* climate crisis
Go (game), 129, 147
Goldwyn, Samuel, 63
Google, 11, 26, 69, 79, 123, 129, 141, 147, 150, 162, 179
Google Home, 141
Google Personalized Search, 150
Gopnik, Adam, 89
Graber, Mauricio, 95–96
Graeber, David, 1, 9
Greenfield, Adam, 104
Griffith, Saul, 25, 52
ground-effect vehicles (GEVs), 181–182
Grundfos, 53, 54
Guardian, 69, 91, 145, 165

H

Harari, Yuval Noah, 94
Harbisson, Neil, 141
Harvard Business Review, 46
Hershey Company, 49
Hill, Bill, 163
Hillis, Danny, 37
Holmes, Oliver Wendell, Sr., 129

INDEX

Horizon Europe, 158–159
Horizon 2020 program, 158
Horizons Canada, 38
Huang, Arthur, 107–108, 118, 120, 127–128
human-centered design, 4–6, 13, 18, 31, 101
humanity-centered design, 13
humans, abilities of AI vs., 129–132, 146
Huxley, Aldous, 45
Hybrid Air Vehicle, 181
hybrid business models, 171–172
hybrid sectors, 162–164

I

"Ice Watch" installation (Eliasson), 58
IDEO, 28
ideological biases, 15–17
IKEA, 117
IKEA Studio app, 143
impact investment, 154–156
Impact the Future initiative, 173
inclusion, proximity expansion for, 76–82
Independent, 91
indigenous design approach, 126
industrialization, 26
"Information wants to be free" conference, 124
Ingels, Bjarke, 75, 139, 188
Ink Hunters, 143
Instant Icon (Skibsted), 51
Institute for Innovation and Public Purpose, 115
Intel, 132
intellectual property (IP), 123–124, 126–127

interaction design, 27
Intergovernmental Panel on Climate Change (IPCC), 58, 110, 138–139
International Space Station (ISS), 78, 163
IP (intellectual property), 123–124, 126–127
IPCC. *see* Intergovernmental Panel on Climate Change
ISS (International Space Station), 78, 163
Italy, 105, 146
Ito, Joichi, 12
Itskov, Dmitry, 91

J

Jacobsen, Arne, 118
Jen, Natasha, 31
Jensen, Søren, 93
Jonkers, Henk, 93
Jorge, Henrique, 91
JUST (company), 99

K

Kartell (brand), 146
Kasparov, Garry, 129, 130, 146
Kennedy, John F., 187
Kenya, 63, 186
Kéré, Francis, 63, 126
Kestrel Materials, 133–134
Kharas, Homi, 111
Kilo, 114
Kintisch, Eli, 139
Kleinberg, Jon, 148
Klynge, Casper, 13
Kolding, Denmark, 103
Kristiansen, Ole Kirk, 41
Kupers, Roland, 172

INDEX

Kurzweil, Ray, 104

L
Lacy, Peter, 112
Larsen, Henning, 142
Lasch, Christopher, 70–71
laser cutters, 65
Lausten, Morten Schwarz, 68
Lefkowitz, Melanie, 148
LEGO, 41, 42, 53, 143, 167
Leke, Acha, 168
Lem, Stanislaw, 1
Lepore, Jill, 12
life-centered design, 31, 80
life expansion, 87–106
   about, 20
   and the anthropocene, 87–89
   applying, 179, 196–197
   artificial life for, 97–100
   for coping with death, 89–90
   digital lives for, 90–92
   living materials for, 92–97
   planet-centered design for, 101–103
   for posthumanity, 104–106
Lincoln, Abraham, 1–2
Linnaeus University, 96
living materials, 92–97
locational utility, 63–65
Loewy, Raymond, 25, 74
London, England, 115
Long Now Foundation, 37
long-term thinking, 36–40, 55, 83
Los Angeles Times, 161

M
Made in Space, 164
Maeda, John, 31
Magna, 46
Mannervik, Ulf, 121, 122
Mars, 139–140, 163, 187–189
Mars Science City, 139–140, 188
Maynard, Andrew, 45
Mazurenko, Roman, 91
Mazzucato, Mariana, 115, 151, 157–158, 162–163
McKibben, Bill, 106
meat, lab-grown, 99–100
Memphis Meats, 99
M/E/P management model, 28
metaverses, 147–150
Microsoft, 125, 170
Milan, Italy, 146
Milan Triennale, 105
Miller, Andrew, 50
MindLab, 73–74
Ministry Root of Causes, 40
Minitel, 14
Miniwiz, 108–109, 118–120, 127
Moedas, Carlos, 158
Mogensen, Børge, 118
Mondragon, 171–172
Monteiro, Mike, 31
Morris, Philip, 108–109
Motorola, 45
Muir, John, 87
multidimensional models of design, 109–110, 127–128, 133–138
multidimensional thinking, 18, 127–128
Muncy, James, 164
Musk, Elon, 88, 140, 160–161, 163, 164
mycelium, 93–94
Mycoworks, 93

INDEX

N

Nanotechnology, 134
Nanus, Burt, 190
NASA, 78, 140, 148, 163
National Public Credit Information Centre, 145
National Science Foundation, 162
Naughton, John, 10
Near Future Laboratory, 47
Netflix, 70
Netherlands, 185
New College (University of Oxford), 33–34, 55
New European Bauhaus initiative, 119, 120
Newson, Marc, 141
New Yorker, 83, 100
Next-Gen (platform), 46
Nike, 108, 113
Nintendo, 142–143
Nipper, Mads, 54
Nordea Foundation, 153
Normann, Richard, 121
Norway, 51, 165
Nuki, Paul, 156
Nussbaum, Bruce, 29, 31

O

Observatory of Public Sector Innovation (OPSI), 160
OECD (Organisation for Economic Co-operation and Development), 160
Ole Kirk Foundation, 41
Oliván, Manuel Castells, 107
O'Neil, Cathy, 78
OpenAI, 88
Opendesk, 66
open-source manufacturing, 66
open-source software, 124–125
OPSI (Observatory of Public Sector Innovation), 160
Organic Basics, 134
Organisation for Economic Co-operation and Development (OECD), 160
Ørsted, 54, 162
Otherlab, 52, 133, 136
Ottesen, Bent, 41
Ovacik, Erdem, 184–185

P

Pacific Rim (film), 98
Page, Larry, 161–162
Paperhouses, 66
Peitersen, Nicolai, 60
Pembroke, Beatrice, 37
Perception (graphics studio), 46
PFA Pension, 152
phage therapy, 14–15
philanthropy, 164–170
philosophy of design, 32
planet-centered design, 101–103
planned obsolescence, 52
Playchess.com, 130
Pokémon Go, 142–143
policy making, 72–75
PoliSpace, 164
Polman, Paul, 35
posthumanity, life expansion for, 104–106
Post-It notes, 28–29, 31
private sector, 21, 43, 152–153, 157–158, 163–164, 169–174
private space tourism, 164
Proctor & Gamble, 29

INDEX

Project Zero, 152-153
proximity expansion, 57-85
    about, 19-20
    applying, 186, 194-195
    democracy for, 72-75
    and digital fabrication, 65-66
    digital proximity for, 67-70
    distance and, 57-59
    ethics for, 59-62
    geography and, 62-63
    for inclusion, 76-82
    and locational utility, 63-65
    people-power gap and, 70-72
    to save Earth, 83-85
public innovation labs, 73-74
public-private partnerships, 152-153
public sector, 152-153, 157, 170-174

Q

Quartz, 96, 168

R

Ramírez, Rafael, 121, 122
Ramo, Joshua Cooper, 64-65
Raworth, Kate, 111
Realdania, 93
recycling, 114, 116, 118-119
Red Cross Experience, 67-69
Re.Designing Death movement, 90
Red Hat, 125
regenerative buildings, 92-93
Replika, 91
reusing products, 119
Ridley, Brent, 134
Rifkin, Jeremy, 111, 170-171
"rights of nature" movement, 80-81
Rivian, 49
Roberts, Debra, 110
robots, 97-100
Rockefeller, John D., 165
Rockefeller Family Fund, 164-165
Rosenberg, Ann, 46
Rosing, Minik, 57
Rosling, Hans, 101
rotorcrafts, 183

S

Salone del Mobile (Milan, Italy), 146
Saltmarshe, Ella, 37
SAP, 46
Savory, Allan, 102
Scandinavian design, 117-118. *see also* Danish design
scenarios, 37-44
Schwab, Klaus, 169
science fiction, 44-47
Scotland, 62
Sea Wolf Express, 182
sector expansion, 151-174
    about, 21
    applying, 179, 187, 198-199
    cross-sector collaboration for, 157-162
    hybrid sectors for, 162-164
    impact investment for, 154-156
    philanthropy for, 164-170
    restructuring private and public sectors for, 170-174
    traditional methods of, 151-154
Se-dol, Lee, 129-130, 147
Selden, Mike, 99
self-driving cars, 79
Shelley, Mary, 44
short-term thinking, 34, 52-53
Sidewalk Labs, 11
Simon, Herbert A., 27

INDEX

Singapore, 99, 186
Singapore Food Agency, 99
Singer, Peter, 57
Singularity, 104
Skibsted, Jens Martin, 6–7, 51, 60, 118, 137
Sloan School of Management, 28
slowing down, 34–37
smart thermal grids, 137–138
Smith, Adam, 67
social credit system, 144–145
social design, 105
SolarCity, 161
Son, Masayoshi, 49
Sønderborg, Denmark, 151–152, 174
South Africa, 63, 165–166
Soviet Union, 14–15, 161, 187
space travel, 187–189
SpaceX, 140, 161, 163
Spain, 66
speed, in design thinking, 30
Spielberg, Steven, 45
Sputnik satellite, 161
Starck, Philippe, 146
StarTAC, 45
Statistics Denmark, 52
Stephen, Zackary, 130–131, 140
StepJockey, 156
Sterling, Bruce, 34, 47
Stern, Nicholas, 22–23
Stevenson, Adlai, 122
Strømann-Andersen, Jakob, 142
Summer, William, 15
sustainable products, 51, 112–114
Swaniker, Fred, 168
Sweden, 96
Switzerland, 114, 117
synthetic biology, 94–95

T

Tataj, Daria, 159
technological determinism, 9–17
   cultural and ideological biases in, 15–17
   resisting, 12–14
   retrospection on, 14–15
   science fiction to combat, 46
   and tech giants, 9–12
Termansen, Anne-Mette, 41–44
Tesla Motors, 161
TGV, 14
thermostats, energy-efficient, 133
3-D printers, 65
3XN, 146
Thunberg, Greta, 84
time expansion, 33–55
   about, 19
   applying, 180, 186, 187, 194–195
   in design fiction, 47–51
   at New College, 33–34
   scenarios for, 37–44
   in science fiction, 44–47
   by slowing down, 34–37
Times Higher Education (website), 168
Toffler, Alvin, 45
topology, 64
Træna, Norway, 51
turbojet train, 14
2045 Initiative, 91

U

Uber, 10
UN. *see* United Nations
UN City (Copenhagen, Denmark), 154
UN Development Programme (UNDP), 159

INDEX

Unilever, 35–36
United Kingdom, 50, 156
United Nations (UN), 54, 81, 95, 103, 110, 122, 135, 153–154, 159, 173–174, 188
United States, public-private partnerships in, 161–162
Universal Robots, 98
University College London, 115
University of Oxford, 33–34, 55
University of St. Gallen, 53
UN Sustainable Development Goals, 54, 81, 95, 103, 135, 153, 154, 173–174, 188
upcycling, 115
Urban Air Mobility, 185–186
urban design, 136
urban mining, 107–110
urban mobility, 184–187
urban planning, 8–9
Ursache, Marius, 91
US Air Force, 143–144
US Army War College, 190
US Department of Veterans Affairs (VA), 1–4, 22, 28
user-centered design, 5, 30–31, 36

V

VA. *see* US Department of Veterans Affairs
value creation, 114–115, 120–123
value expansion, 107–128
    about, 20
    in anthropocene epoch, 110–111
    applying, 186, 187, 196–197
    circular economy for, 112–120
    multidimensionality for, 127–128

    redefining value protection for, 123–127
    urban mining for, 107–110
    value-creating networks for, 120–123
value protection, 123–127
van der Poel, Peter, 117
V&A Publishing, 97
Verganti, Roberto, 31
Verge, 91
Verne, Jules, 44–45
Vestas Wind Systems, 162
virtual reality, 67–69, 142–144
VOI (firm), 185
Volkswagen, 49
von der Leyen, Ursula, 119

W

Walters, Helen, 29, 31
Waterfront Toronto, 11
wearable technology, 141
WEF. *see* World Economic Forum
Wegner, Hans J., 118
wetware computers, 97
Wohlleben, Peter, 102
World Economic Forum (WEF), 114, 117, 137, 169–171
Wylie, 11

X

Xi Jinping, 50

Y

Yonhap (news agency), 147

Z

Zipline, 63

ABOUT THE AUTHORS

Photo by Rainer Hosch

Jens Martin Skibsted, partner at Manyone, is a multiple award-winning designer, entrepreneur, and design philosopher. Best known for his urban mobility designs for Biomega and collaborations with design superstars such as Marc Newson and Bjarke Ingels, he helped found several design consultancies including Skibsted Ideation, KiBiSi, and Manyone. His designs live in the collections at the MoMA, Le Cnap, Designmuseum Danmark, SFMOMA, and more. He is on the boards of the Danfoss Foundation, Biomega, and Manyone, and is a member of the World Economic Forum's Global Future Council on Mobility Transitions. He is formerly the chairman of the Danish Design Council, the vice chair of the World Economic Forum's Global Agenda Council on Design Innovation, and co-chair of the World Economic Forum's Global Future Council on Entrepreneurship, and served on the advisory board of the

INDEX prize and MindLab. Jens Martin is also a Young Global Leader alumnus who has spoken at Davos, DLD, TED, and more. His writing has appeared in the *Huffington Post*, *Harvard Business Review*, *Børsen*, *Washington Post*, and *Fast Company*.

Photo by Agnete Schlichtkrull

Christian Bason is CEO of the Danish Design Centre, a nonprofit foundation backed by the government to advance the value of design for business, society, and the planet. He is former director of MindLab, the Danish government's innovation team, and former business manager at Ramboll, a global consultancy. He is the author of numerous books on innovation, design, and leadership, including *Leading Public Design* (Policy Press, 2017) and *Design for Policy* (Routledge, 2014). He has written for, among others, *Harvard Business Review* and *Stanford Social Innovation Review*, and is a columnist with the Danish weekly *Monday Morning*. Christian is a board member of the Royal Danish Academy, the Rockwool Foundation Intervention Committee, and CfL Centre for Leadership; a member of the World Economic Forum's Global Future Council on Agile Governance; and former chairman of the European Commission's Expert Group on Public Sector Innovation. He is a frequent keynote speaker and an executive lecturer with Henley MBA, Copenhagen Business School, and the European School of Administration. Christian earned his MSc in political science and PhD in design leadership. He was honored in 2018 by German daily *Tagesspiegel* with their inaugural Creative Bureaucrat Award.